MASS CUSTOMIZATION

OPPORTUNITIES, METHODS, AND CHALLENGES FOR MANUFACTURERS

Hans Kull

Apress®

Mass Customization: Opportunities, Methods, and Challenges for Manufacturers

Copyright © 2015 by **Hans Kull**

This work is subject to copyright. All rights are reserved by the Publisher, whether the whole or part of the material is concerned, specifically the rights of translation, reprinting, reuse of illustrations, recitation, broadcasting, reproduction on microfilms or in any other physical way, and transmission or information storage and retrieval, electronic adaptation, computer software, or by similar or dissimilar methodology now known or hereafter developed. Exempted from this legal reservation are brief excerpts in connection with reviews or scholarly analysis or material supplied specifically for the purpose of being entered and executed on a computer system, for exclusive use by the purchaser of the work. Duplication of this publication or parts thereof is permitted only under the provisions of the Copyright Law of the Publisher's location, in its current version, and permission for use must always be obtained from Springer. Permissions for use may be obtained through RightsLink at the Copyright Clearance Center. Violations are liable to prosecution under the respective Copyright Law.

ISBN-13 (pbk): 978-1-4842-1008-6

ISBN-13 (electronic): 978-1-4842-1007-9

Trademarked names, logos, and images may appear in this book. Rather than use a trademark symbol with every occurrence of a trademarked name, logo, or image we use the names, logos, and images only in an editorial fashion and to the benefit of the trademark owner, with no intention of infringement of the trademark.

The use in this publication of trade names, trademarks, service marks, and similar terms, even if they are not identified as such, is not to be taken as an expression of opinion as to whether or not they are subject to proprietary rights.

While the advice and information in this book are believed to be true and accurate at the date of publication, neither the authors nor the editors nor the publisher can accept any legal responsibility for any errors or omissions that may be made. The publisher makes no warranty, express or implied, with respect to the material contained herein.

>   Managing Director: Welmoed Spahr
>   Acquisitions Editor: Robert Hutchinson
>   Developmental Editor: Douglas Pundick
>   Editorial Board: Steve Anglin, Mark Beckner, Gary Cornell, Louise Corrigan, James DeWolf,
>   Jonathan Gennick, Robert Hutchinson, Michelle Lowman, James Markham,
>   Susan McDermott, Matthew Moodie, Jeffrey Pepper, Douglas Pundick,
>   Ben Renow-Clarke, Gwenan Spearing, Matt Wade, Steve Weiss
>   Coordinating Editor: Rita Fernando
>   Copy Editor: Kim Wimpsett
>   Compositor: SPi Global
>   Indexer: SPi Global

Distributed to the book trade worldwide by Springer Science+Business Media New York, 233 Spring Street, 6th Floor, New York, NY 10013. Phone 1-800-SPRINGER, fax (201) 348-4505, e-mail orders-ny@springer-sbm.com, or visit www.springeronline.com. Apress Media, LLC is a California LLC and the sole member (owner) is Springer Science + Business Media Finance Inc (SSBM Finance Inc). SSBM Finance Inc is a Delaware corporation.

For information on translations, please e-mail rights@apress.com, or visit www.apress.com.

Apress and friends of ED books may be purchased in bulk for academic, corporate, or promotional use. eBook versions and licenses are also available for most titles. For more information, reference our Special Bulk Sales–eBook Licensing web page at www.apress.com/bulk-sales.

Any source code or other supplementary materials referenced by the author in this text is available to readers at www.apress.com. For detailed information about how to locate your book's source code, go to www.apress.com/source-code/.

Apress Business: The Unbiased Source of Business Information

Apress business books provide essential information and practical advice, each written for practitioners by recognized experts. Busy managers and professionals in all areas of the business world—and at all levels of technical sophistication—look to our books for the actionable ideas and tools they need to solve problems, update and enhance their professional skills, make their work lives easier, and capitalize on opportunity.

Whatever the topic on the business spectrum—entrepreneurship, finance, sales, marketing, management, regulation, information technology, among others—Apress has been praised for providing the objective information and unbiased advice you need to excel in your daily work life. Our authors have no axes to grind; they understand they have one job only—to deliver up-to-date, accurate information simply, concisely, and with deep insight that addresses the real needs of our readers.

It is increasingly hard to find information—whether in the news media, on the Internet, and now all too often in books—that is even-handed and has your best interests at heart. We therefore hope that you enjoy this book, which has been carefully crafted to meet our standards of quality and unbiased coverage.

We are always interested in your feedback or ideas for new titles. Perhaps you'd even like to write a book yourself. Whatever the case, reach out to us at editorial@apress.com and an editor will respond swiftly. Incidentally, at the back of this book, you will find a list of useful related titles. Please visit us at www.apress.com to sign up for newsletters and discounts on future purchases.

The Apress Business Team

To Anne, Judith, Sonia, and Simon.

Contents

About the Author. ix
Acknowledgments. xi
Preface. xiii

Chapter 1: Introduction 1

Part I: Horizontal Integration in Mass Customization. . . 7
Chapter 2: Intelligent Manufacturing Technologies 9
Chapter 3: Problems in Smart Factory Projects 21
Chapter 4: Future Challenges 35

Part II: New and Emerging Technologies. 41
Chapter 5: New ICT Technologies. 45
Chapter 6: Alternative Factory Floor Interface Devices. 55
Chapter 7: New Manufacturing Technologies. 59

Part III: Vertical Integration in Mass Customization. 71
Chapter 8: Software Integration 73
Chapter 9: Going Cloud 77

Part IV: Making Mass Customization Happen 91
Chapter 10: Optimizing System Performance. 93
Chapter 11: Commissioning. 109
Chapter 12: Implementation Process. 117
Chapter 13: References 125

Index 129

About the Author

Hans Kull is the managing director of inmatic in Australia, which develops software for manufacturers, particularly mass customization projects. He was formerly the owner of Kull Informatik and before that head of the systems department at Kaba AG in Switzerland. Throughout his career he has worked on software projects in the manufacturing industry, most of them involving combinatorial optimization and process automation. He was an early adopter of object-oriented programming and agile software development. He studied mathematics at the ETH Zürich (Swiss Federal Institute of Technology), achieving a master's degree and PhD, and before that he studied electrical engineering at the ZHAW (Zurich University of Applied Sciences), where he was a mathematics lecturer for three years as well. He is a member of ACM, IEEE, and the IEEE Computer Society.

Acknowledgments

Many thanks go to Glen Carlson and the whole team of KPI Australia for helping me with the start of this journey. A big thank-you goes to my family and all my employees for putting up with me during all my ups and downs in the process of writing this book.

Furthermore, I would like to thank in particular Morgan Beale for reading my first version of the book and for all his helpful comments and improvements.

Last but not least, I would like to thank the whole team at Apress for looking after me so well during the process: thanks go in particular to Robert Hutchinson for accepting my manuscript and proposing it to the board, as well as for all the help with improving its structure; to Rita Fernando for helping me through the editing process and answering all my questions; to Douglas Pundick for checking all the technical details; and to Kim Wimpsett for that great edit of my copy.

Preface

Mass customization is not new. There was a bit of a buzz about it, but in the past 10 to 15 years it has become a bit quieter. However, I think the time has come to talk about mass customization again.

This is because a lot has happened in that time in the field of information and communications technology (ICT) as well as in manufacturing technologies. At my company inmatic, the projects we do have evolved more and more from smart factory to mass customization projects. Some projects were new, and others were upgrades of existing factories with new machines and software to make mass customization happen to the fullest.

And it doesn't stop here. With new developments such as cloud computing and the Internet of Things (IoT), as well as new manufacturing technologies such as additive manufacturing and smart robotics, not only manufacturing becomes smarter but products and components become smarter as well. There is a huge potential in going much further in the direction of mass customization in existing factories by extending the degree of automation as well as by extending mass customization into a much wider range of industries.

This book is about all that. It is a book for manufacturing executives who want to see the opportunities and challenges in mass customization. And of course you want to learn about the problems, the methods, and the risks and how to mitigate them. At inmatic, I have been lucky to work for industries where mass customization is possible for quite some time now, and in the past ten years, having managed about a dozen such projects and developed part of the software for them, I would love to share my experiences with you.

CHAPTER 1

Introduction

"A pessimist sees the problem in every opportunity; the optimist sees the opportunity in every problem."

—Winston Churchill

In the manufacturing world, *mass customization* means the automated manufacturing of bespoke products. For example, windows come in all sizes and shapes but are produced with the same machinery in large quantities. Mass customization can be understood as the combination of the concepts of *direct digital manufacturing* (DDM) and the *smart factory*.

DDM happens along the vertical integration path, whereby the design of a product is passed on in digital form to the suppliers, be it in total or in components, and each supplier adds its specific data in order to produce its component. In the end, the data is passed on to the machines, and the parts are produced directly from that data.

Smart factories build products automatically in larger quantities. This happens along the horizontal integration path, where machines produce and pass on parts and components until they are finally brought together on an assembly line. Either they know themselves what and where the different components are or, in a more flexible production environment, components are able to identify themselves and give all necessary information—for example, to the assembly line.

In 2012 the German government started the Industry 4.0 initiative. It talks a lot about those issues. I will describe this initiative in more detail in the next chapter. When I read the report on Industry 4.0 [1], I noted that they were to a large extent talking about what my company inmatic has been doing for a long time. Of course, there is more to Industry 4.0 than what we do, and yes, what we (and others) have done could until recently be done in only a few specific industries. Industry 4.0 is about the new and emerging technologies that will enable mass customization for a much wider range of industries.

That's what this book is about. Admittedly, there is more to Industry 4.0, which I will mention later, but that isn't central to this book.

My company inmatic has more than ten years of experience with mass customization projects, with me in the leading role on most of them. In fact, my first big project more than 30 years ago was at the core of a mass customization project. I will talk about some of those projects later. These projects did not automate each and every aspect of order and production processing. However, I have seen some of those aspects evolving over a longer period of time, and some of the productivity gains are staggering. Here, I will pass on my experiences with these mass customization projects, despite that this idea is completely new for most industries.

But this book is not about 35 years of experience as an engineer and mathematician in information and communications technology (ICT). It is about the opportunities provided by new and future technologies and how you could use them in your projects. When I talk about the past, I do so to give an example of how things could be done and what could be improved. When I talk about the lessons learned, I do so only to explain why I would do these things in the way that I suggest.

To illustrate what I mean by mass customization, consider the following challenge: how could you produce custom stairs for houses in a completely automated way, including design, construction, and administration?

Our stairs manufacturer for this example will have their standard ways of constructing and producing stairs. On their website will be all the types of stairs you could think of for customers to choose.

Let's say you have added a second story to your house, and now you need something nicer than a ladder to go upstairs. You know where it has to go and have taken exact measurements. So, you go to this web site and pick the type of stairs you need, something going up about 1,350mm (4.43') to a landing and then turning left and going up to the first floor at 2,687mm (8.82').

After entering the basic measurements, you enter the adjustments needed; for example, the walls might not be exactly at right angles. You will specify whether you want a post in the corner or whether you have walls, and you can specify the type of the hand rails and where they have to go. Of course, you can also specify the material of the stairs themselves, so for example, if this is a timber stairs manufacturer, you can select the type of timber you want to use as well as its treatment.

Now the software on the server is able to calculate all the parts needed to build your stairs, together with their exact measurements and all of the machining, processing, and treatment. It can produce a drawing with all the measurements for you to verify. The system is also able to calculate manufacturing costs and therefore gives you an immediate quote. If you ask for delivery to your home and give the address, then it can calculate the transport distance and thus delivery costs. And if you ask for installation by the manufacturer, then it can quote for that too.

This means you are now able to order. You might compare a few other manufacturers, and then once you have made your decision, you go back to the web site, log in, go to your quote, order, and pay a deposit.

Your order triggers a lot of data processing in the background. Your job will be scheduled, delivery confirmed, machine code generated, and automatic production planned. Once your job is up for production, the machine code is loaded onto the machines, and all the parts are produced and automatically delivered to the assembly line, where most likely your stairs are preassembled to the parts that will be installed, for example the corner post first, then the landing, and finally the two stairs segments.

I hope this illustrates how mass customization could work. In fact, all the technologies for this to happen are available, and we have done similar things already with cabinets.

What is new and exciting is that some key enabling manufacturing and ICT technologies have become available that, now and in the near future, enormously widen the range and type of products that can be mass customized. Later in this book I will talk about these technologies in a bit more depth.

But ICT is not just computers and networks and software; it is about humans too. Experiences and lessons learned are important, which is why I will share some of mine. We humans do not change at the same relentless rate as technology does now. So, I will talk about humans in that environment. I will talk about people and their roles in ICT, about chief information officers (CIOs), project managers, architects, and programmers. I will talk about company directors and chief executive officers (CEOs), and I will talk about the workers in the offices and on the factory floor who will have to cope with all that change.

New investments mean new risks. The world seems to change at an ever-faster pace. Smart factory, advanced manufacturing, customized products, flexible systems, digital manufacturing, agile, the next Industrial Revolution, Industry 4.0, the Internet of Things...the industry is humming with buzzwords, but nobody really knows what is coming. Manufacturing has millions of different faces, but they all have one thing in common: new ICT technologies will change them all.

In an ever-changing environment, it becomes harder and harder to understand those risks, let alone to mitigate them. Higher degrees of automation mean more complexity. This in turn means longer startup times for the new machinery, more problems and hiccups, and a longer period in which you are not able to satisfy your customer's needs. What if the machinery takes too long to come up to speed? What if it does not perform as specified? What if it does not do what is needed? What if you ordered the wrong thing? What if the market changes in a year or two but the new machinery can't change production to match?

This last point needs a bit more attention. The question is no longer "What if the market changes?" Instead, it is "How will the market change?" More and more business is done online. Most likely this will affect you too. If you are in a situation where you must and can invest, this is not only a threat; it can be an opportunity too. This might be your chance to define how the market will change in your industry. If you see and understand the new and coming technologies, you can work out how you can do business better. This book will help you gain that insight.

This book is about sophisticated uses of ICT in manufacturing. Specifically it's about order processing, automation, integration of business processes and production processes, and optimization of all those processes; in short, it's about using ICT to produce as efficiently as possible with as few resources as possible. It is about connecting your order processing system with production planning, with your machine control and management, and with your dispatch system. It is about integrating your whole factory into a single cyber physical production system (CPPS).

This is only one part of smart manufacturing, though. The other part is smart products and components. If your machines and your manufacturing execution system (MES) can ask the parts what they are, then this opens a whole range of possibilities for your manufacturing. However, this will not always be possible or make sense. As long as you make sure all the screws in the magazine are the same, it will not be necessary that every screw is able to identify itself and give its specification.

Smart manufacturing will affect all aspects of production. Figure 1-1 shows the main ideas of mass customization. It follows an idea from [2], modified and adapted for our purposes.

Mass Customization

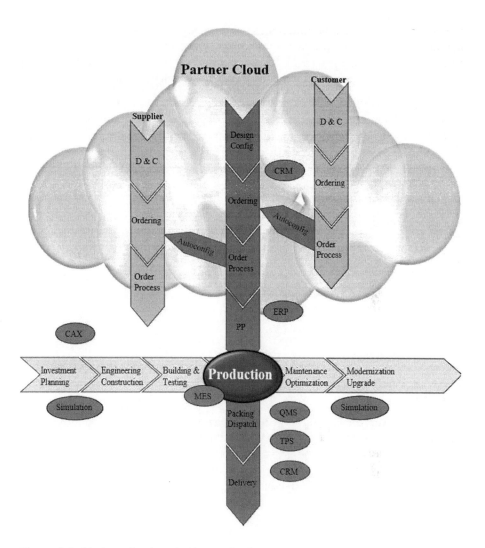

Figure 1-1. Horizontal and vertical integration in mass customization

Figure 1-1 shows that in the future mass customization starts in the cloud. You can't do all the customizations yourself, so part of what is drawn in the cloud will have public access if you supply to consumers. It's the place where they can design and configure products to fit their needs and where they can order those products.

For higher-tech products, you might want to restrict access to your knowledgeable customers only. This will usually be the case if you deal mainly with business-to-business (B2B) sales. But even then you might give public access for simple things such as spare parts and the like.

During order processing, you might have to order in external parts. This means that some of your suppliers might have to mass customize too. If you have a defined interface to your suppliers, then this can be automated so that your orders are released and processed automatically on your supplier's side. I will give examples for this.

At my company inmatic, we have been involved with all of these aspects of mass customization in some way or other. We have developed software along the order processing track from customer relationship management (CRM) and order entry to production planning and production management and control, packaging, and dispatch.

As systems integrator we have been involved with the factory planning and engineering and in the same capacity in researching and implementing new technologies, particularly for manufacturing machinery and support.

In that sense at inmatic we are always aware of new technologies and in particular how ICT has changed products and production as well as data processing and administration. This book is about my experiences and visions for these new technologies.

There are four parts to this book. The first helps you better understand problems you might have encountered in recent automation projects. It talks about what is different in manufacturing projects now, compared to only a few years ago. It talks about my experiences and the problems I have encountered as well as my vision for handling these projects in the future.

In the second part, I talk about new technologies in ICT and about manufacturing-related ICT technologies. I talk about what is already available, what is just round the corner, and some technologies that will take a while longer. The segment is about opportunities and gives a vision of how you could improve your manufacturing facilities beyond what machinery manufacturers and software houses offer.

The third part is about opening to the outside world. It is about extending the network and about partnerships and collaboration with customers and suppliers and what that means in terms of ICT. It is about how to make sure all data is entered only once. It is about the supply chain as well as after-sales support and a little bit about the product life cycle as well.

The last part is about what you can do once you have crossed the new investment speed bump and how you could further improve your factory. It discusses what investments, small and a bit larger, you could make to further improve your manufacturing site—be it by increasing productivity, reducing waste or saving energy, or all three. Finally, it summarizes mass customization and what it could mean to you.

PART

I

Horizontal Integration in Mass Customization

"In any moment of decision, the best thing you can do is the right thing. The worst thing you can do is nothing."

—Theodore Roosevelt

By horizontal integration I mean the integration of machinery so that parts and components are automatically passed from machine to machine and the machines know either from the manufacturing execution system or from the part itself what they have to do with those parts or components.

Mass customization, smart factory, and Industry 4.0 all mean similar things: essentially the increased use of information and communications technology (ICT) in the manufacturing process. In this scenario, only the factory is smart, not the products or parts themselves. The term advanced manufacturing is mainly used and occasionally recycled in English-speaking countries. It is currently used in a similar context as smart factories, but in addition it refers to other new and emerging manufacturing technologies.

In this part, I discuss these concepts and the problems and challenges that come with them. In Chapter 2, I explain the terminologies and concepts. At the end, I introduce two examples where mass customization has been happening for a long time already. I later refer to and extend these examples to give a more in-depth understanding of the problems that arise in mass customization projects.

Chapter 3 explains how problems from software development projects now spill over into mass customization projects, as almost all the additional complexity that comes with mass customization arises in the software of such projects. But although the proportion of software as part of the overall cost will rise, software will still not be the dominating cost factor, a fact that can make things even worse. This because most complexity is in the software and needs more attention than what its proportion of the overall costs would imply.

Chapter 4 is about future challenges that come with all the new technologies that are available now or will be in the near future.

CHAPTER 2

Intelligent Manufacturing Technologies

> *"We are stuck with technology when what we really want is just stuff that works."*
>
> —Douglas Adams

The Smart Factory

The term *smart factory* is old now. Most of the automation projects that I've done in the last ten years could be seen as smart factories. Many of the examples mentioned in the final report of the Industry 4.0 working group [1] are that type of smart automation.

At the end of this chapter, I will explain how the production of *insulation glass units* (IGUs) is automated. But let's see first how the IGU manufacturer gets his data. The glass manufacturer gets his orders mostly from window manufacturers, who in turn get their orders from architects. The architect can simply export her data and order the windows, adding some specifications regarding insulation properties, wind loads, and so on. The window manufacturer can then build on those drawing extracts and add the information he needs to be able to produce the frames and order the IGU.

Next, the glass manufacturer gets the specifications of the IGUs and transforms requirements into information; for example, required wind load becomes a specified glass thickness. Once these specifications are finalized, the data can flow into the factory.

An essential point is the fact that no data has to be entered twice. In the ideal world of smart window manufacturing, an architect designs the size and specifications of the windows in her drawings and orders them electronically. The window manufacturer extracts all data from the drawings automatically and specifies the frame profiles. The sizes and specifications of the IGUs come from that and are ordered electronically with the glass manufacturer. From this specification, the glass types are determined, and the units go into production batches. For these batches the different pieces go into layouts, are optimized and cut, and then go automatically to assembly.

At each stage along the supply chain some additional data is added, but the initial data is never entered again. It is only complemented with additional information as needed. See Figure 2-1.

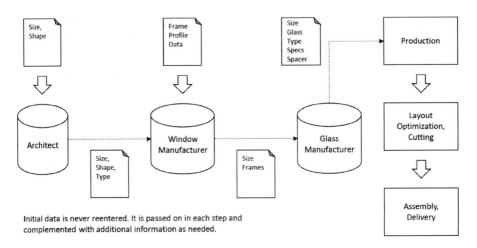

Figure 2-1. Data flow in window manufacturing

As with the window manufacturer who processes the frames himself and orders IGUs externally, the manufacturer of the IGUs might have to order parts externally. For example, if laminated or toughened glass is required, he orders that from another specialized company. Once all orders are confirmed and a delivery date given, the production of the IGUs can be planned, and thus a delivery date confirmed and sent back to the window manufacturer, and so forth, until the building manager receives an order confirmation and an installation date for the windows.

Mass Customization

What comes next is a stream of material up the chain following the order confirmation. If everything works out as planned, then the toughened glass is delivered on time for assembly to the IGU manufacturer. From there, the IGU is delivered on the confirmed date to the window manufacturer, and they install on the confirmed date.

You certainly have heard about all this before and might wonder what is so smart about it. Car manufacturers have produced in this way for decades. Well, what is new is that what you can produce is very individual. Yes, when you order your new car, you have a few choices. You can specify a basic version or a higher-end version of the model you picked and with that selection comes a different set of features to pick from. Thus, you can specify color and a few accessories. Some of these features are installed at the local dealer rather than the car factory, but there are still a few thousand combinations of options for the car manufacturer.

So, where is the difference? At the assembly line of the car manufacturer there is always the same make of a car assembled. Yes, the options mean that, for example, the robot that installs the dashboard installs different dashboards for different models. But there is no need to reprogram that robot. It always makes the same movement. All that is needed is to ensure that when each car body arrives, the right dashboard is there for the robot to pick up.

In a smart factory, everything is different. It is not as simple as just specifying the size and quality of a window and all processes flowing from that. Down the ordering chain information has to be continually added.

Not all windows are rectangular. And on top of complex shapes, there can be cutouts, for example for hinges if you have all-glass windows or doors. And because the windows can all be different, the machinery has to do different things all the time. The glass cutter has to cut different shapes and sizes, the sorter has to store different sizes, and the sealing line has to produce units of different shapes and sizes as well. Of course, this happens upstream at the toughened glass manufacturer as well and again downstream at the window manufacturer's site.

The difference is at the production level. With the car manufacturer, the production line always does essentially the same thing. A bit of outside management just makes sure there are always the right parts at the right place for assembly. Don't get me wrong. This can be complicated enough. But in a smart factory, the production differences go to the parts level, and the assembly line has to do very different things all the time, down to a batch of size one.

As an example, the Industry 4.0 report points out that with the advancements of laser sintering technology it is now possible to produce various materials, such as alumina, stainless steel, titanium, and many plastics, with additive manufacturing methods. This means an engineer at one company can design a part using a computer-aided drawing (CAD) tool and then send the design to another company where it is 3D printed. This is the smart factory concept. Machines produce individual pieces in a profitable way. All the intelligence is in the machines and in preprocessing machining data for them. The parts themselves do not hold any production-related intelligence.

The Promise of Industry 4.0

Industry 4.0 is currently not much more than an idea. Not many industries can produce individual items in a completely automated way. For this to happen, not only the machines but sometimes even the parts themselves have to become smart.

The German white paper on Industry 4.0 [1] goes a bit further than the concept of the smart factory by bringing in the idea of using the Internet of Things (IoT) in the production cycle. There are other ideas around those concepts as well, for example cooperative design and construction in the cloud, as well as handling the supply chain in the cloud. The authors of the Industry 4.0 report point out that this scenario is still about 15 to 20 years away. I'm of the opinion that a lot of those ideas could already work in many industries.

I will include a few examples of new technologies that might well become enablers of mass customization in a wider range of industries. That is particularly likely where data can be generated and sent to a production machine with little or no human intervention so that wide aspects of the production of custom products can be automated.

It all depends on what you want to produce. In a smart factory, the manufacturing execution system (MES) controls the production operations. Typically it knows which part is where, where that part has to go next, and what has to happen to it once it is there. Those parts do not have to be smart at all.

In smart factories, data from order processing goes directly to the machines. Depending on the complexity of the processes involved and how they are connected, a simple approach to optimize production might suffice, where in other cases a smarter approach might be needed.

As you will see in the following examples, in certain industries this allows automatic production of items that, in extreme cases, are all different. IGUs can have different sizes and shapes, and locks and keys can have different codes.

If not only the machines have become smarter but the parts as well and if they can communicate effectively, then the automated production of more complex customized products becomes possible. This extended potential of horizontal integration is one key aspect of Industry 4.0. It will be one enabler of mass customization in a wider range of industries. I will talk about vertical integration, another enabler and crucial part for mass customization, in Part III of this book.

As described in the example about the IoT, it is helpful when the parts can identify themselves. The standard for current operations in my machine projects is that external parts come with a barcode label identifying them. This has the disadvantage that upon delivery all parts have to be scanned. Only those that can be processed in the near future, though, can be sorted directly into the MES. The others have to remain on their racks. If the planned production time is known, then the user is given that information so she knows where to store the rack and can find them again when they are needed.

This solution has the disadvantage that you can lose track of the pieces that are not sorted in. When they are later needed, you can't tell the user where he can find them.

In the Industry 4.0 demo factory that the German Research Centre for Artificial Intelligence (DFKI) has built in cooperation with 20 partners, they fill soap bottles [3]. The empty bottles get a radio-frequency identification (RFID) tag. The tag can then tell the machines whether they should get a black or white cap.

To me this seems not to be a very smart bottle, though. The reason this sample was used lies in the fact that RFID tags have become very cheap. But this is mainly true for the simplest version of them, and they don't know more than what their name says, in other words, their ID. This is OK; it just means that the MES then has to know more about that part. And that could have been achieved by a barcode, provided it could have been automatically read on the machines.

However, the main difference under Industry 4.0 is that the parts themselves have some form of intelligence and are able to communicate with the outside world. So, you have a smart factory as described earlier, networked with customers and suppliers, and in addition to that the products and parts are themselves smart.

I'm the first to admit that not all the products of the world can be built in smart factories the way I described in this chapter and Chapter 1. I was just in the privileged situation to come across types of products where that became possible. So, in many cases, it makes sense if the parts and in the end the whole product becomes smart as well.

Imagine you want to build a fairly complex electronic system, for example a laptop computer. Imagine there are dozens of options, including the color of the housing, the type and resolution of the screen, the language of the keyboard and the operating system, the type and size of the hard disk, the memory speed and size, the processor type and speed, the software to be installed, and more.

When ordering online, the customer can pick each of these features, like on the Dell web site. Imagine everything happening automatically after the payment is processed. The specification part of the order enters the MES. This first writes some type of IoT chip (could be RFID) and loads it with all the information about the production of your laptop computer. And with that chip the production process starts. The first machine reads the chip and understands that you would like a silver housing. It grabs that from its store, attaches the IoT chip to it, and sends it up the line where at the next station the appropriate motherboard is installed. This then continues on; your laptop travels up the assembly line, stops at the appropriate places, and gets what it needs according to the IoT chip's data.

This production method would have a lot of advantages. For a start, it saves connecting all of the machines to the central computer on which the MES runs. The MES no longer has to keep track of all the parts, although it might be interested in the progress, which means that at some points of the assembly line, the IoT chips report back.

One big advantage is that the system is robust against failures. Let's for example think of a situation where at some station the operation fails or can't be performed because of a missing part. In that case, that partial product is simply removed from the line, either by the operator or automatically. There is no confusion of operation anywhere on the assembly line because every product knows from the start what it wants to become. And once the problem of the removed partial product is fixed, it can simply be returned to the assembly line somewhere before the point where it was removed.

Smart parts and products know something about themselves and something about their production too. They can control their assembly semi-automatically.

In Industry 4.0, these smart products remain smart throughout their productive life. They know something about their maintenance needs and will have a way to communicate them to the outside world. Here the Internet of Things comes into play. In the near future you will have to register your new dishwasher with your Wi-Fi net so that it can communicate with the outside world. If something is not right anymore, you will get a message on your tablet or your smartphone telling you what the problem is. If you allow it to connect to the Internet, then it will contact the next repair center, and they will know what to do. Or maybe, if you have your own 3D printer or know some additive manufacturing business nearby, it will just provide you with the data to print the damaged part and a manual to replace it.

And should you at some point decide that your dishwasher is now past its use-by date, then at the recyclers it will tell them what's inside to salvage and where—provided your smart dishwasher is still smart enough when it arrives there.

Mass Customization Types

Early examples and a lot of early literature on mass customization are configuration-type mass customizations. A famous example is Dell, which allowed consumers from the mid-1980s to configure the components of their computers to their needs. A more modern example is BMW with the Mini Cooper, where you can pick and choose from some 100 options to configure your car. Some people tend to see this type more as a customized assembly and not as true mass customization. As long as the choices are easy to verify, for example because there are just mutually exclusive sets of properties, and as long as the assembly in the end is essentially manual, I can understand where they are coming from.

In contrast to this stands the parameterized type of mass customization, where you can change, for example, the size and shape of a window or the size of a cabinet and, for example, the number and height of drawers for a cabinet.

Both types of mass customization have their problems and complexities. For example, when personal computers were still a big box under your desk, it was much easier to accommodate all your configuration wishes for a PC. It might have meant a bigger power supply at some point, but there was almost always enough room to accommodate all the components you wanted. With laptop computers came many more restrictions, and configurations became harder. Apart from the problems with space, you can't just increase power; you would have to get rid of the heat as well, and battery life becomes an issue too.

Such complexities arise in a parameterized environment as well. For example, if a window becomes big, you might have to increase the glass thickness to carry the wind load. This in turn might mean making the frame sturdier to carry the heavy insulation glass unit and increasing the type and number of hinges for the same reason.

In the future, you will see more and more that these types of mass customization will merge; that is, you will see more and more applications of mass customization with parameterization aspects as well as configuration aspects within them. This of course will increase the level of complexity that has to be handled even more.

Prototype Case Studies

I start here with two examples that will be described throughout the book. I have worked on these two projects—or, better, *processes*—over long periods of time. That is, I started on these processes at some point and over the years added more and more functionality. And it was not only me, there were my programmers at inmatic involved and sometimes the customers and the machinery manufacturers. Functionality was added to the software, machinery, and automation process until reaching a state of manufacturing that can be called mass customization in its full meaning.

All mass customization projects have to start on the vertical integration part. You need at least for the most common part of production a way to specify the products as efficiently as possible. And you need the specification of the products in digital form, before you can start any automation process.

This would mean that you should start the book with the vertical integration coverage rather than with the horizontal one. But these examples start in a time before PCs even existed, and I do not want to go on about the olden times more than necessary. In Part III on vertical integration, I will talk about what you can do now and in the near future.

Locks and Keys

One of my first employers after I finished my math studies was a manufacturer of lock cylinders and keys in Switzerland. These days, they do more in the field of access control than that, but back then it was almost all about mechanical lock cylinders and keys. For those not familiar with the terminology, a *lock cylinder* is a part that is inserted into the lock. Its function is to decode the key, and if the code on the key is correct, the lock cylinder allows the key to turn and thus open or close the lock, in other words, to drive the bolt in the lock.

Larger buildings have specific access requirements. For example, in an apartment block, each tenant should be able to open the front door and the door to his apartment only. Maybe he should be able to open his mailbox as well and the door to the underground garage all with the same key. In office buildings, factories, or hospitals, for example, these requirements can become complex.

Customers would use a large spreadsheet, with all the keys in the first column and all the doors in the top row, and then indicate with an *x* in the intersection field if a key had to open a door. Such a system of locks, keys, and opening functions is called a *master key system*. (The key that opens all the doors is called the *master key*, so I guess that's where the name comes from.)

Back then, calculating the codes for the cylinders and the locks was done manually. To some extent there was a system to it. But that was not enough; somehow this was an art form as well. Requirements could become complex. In contrast, in a mechanical system, the numbers of pins you can play with is limited. So, it sometimes took two weeks of analysis and calculation just to figure out that a customer's requirements could not be fulfilled.

Essentially I got this job because the company wanted to computerize this calculation. It was not easy to make this work, and I will talk about the problems encountered later in this book. What is relevant here is that it was clear to

Mass Customization

everybody that if we could generate and manage the master key system on the computer and store all the codes of the cylinders and keys, then a lot of automation could flow from that.

Once codes were available on the computer, we could send them to the key cutting machines as well. These were already microprocessor controlled. People had to enter codes on a keyboard on the machine. Now we could send the codes to the machines where they were cut automatically. Assembling cylinders automatically was a bit more complex, and it took a while until that worked as well. We started by first printing the codes and later displaying them on screens at the manual assembly stations instead. Then gradually the assembly of the various products and types of cylinders was automated.

This led to massive reductions in delivery times, which gave the company a huge competitive advantage. Even in the first year after introducing the new system, when cylinders were still assembled manually, annual sales went from about 28 million to 44 million.

So, in the mid-1980s, the company started implementing a smart factory, completing it ten years later in the 1990s, even though *smart factory* wasn't coined as a phrase until the early 2000s. That is, with little human effort, the data was flowing from order entry directly into production, and keys were cut and cylinders assembled from that data automatically, again with little human intervention. And the company reaped impressive rewards for doing so.

In their most recent development, this manufacturer has now started to produce inserts that contain only the decoding part of a cylinder. These inserts in turn fit into different housings to make the different types (in other words, building forms) of lock cylinders.

The outer shape of these inserts is always the same, which makes it much easier to build a completely automated production line. The obvious next step is to automate the assembly of inserts into housings. The tricky bit now is to make sure the right inserts go into the right housings.

To explain that a bit further, remember how the customer provides his specifications. For every door he gives a name for a lock cylinder and then indicates which keys should open that door. This means every lock function has a name given by the customer. This name is written on the cylinder (now the housing), together with the master key system identifier. This in turn allows the locksmith to identify the cylinder and install it at the right door.

But now the inserts do the decoding, so they have to carry that identification too. If the company could make sure that they produced the inserts in exactly the sequence they were used when assembling them into housings, then that would not be necessary. But imagine what would happen if something goes wrong with one insert and it falls off the production line or has to be removed. They would no longer know which insert is which, and even if they did, they would have a hard time bringing the insert production line and the cylinder assembly line back into sync again.

On the other hand, if the inserts could identify themselves at the cylinder assembly line, then that line could react automatically if an insert goes missing. All it has to do is back order the missing insert at the insert production line and postpone the assembly of that cylinder until the replacement insert arrives.

There are many ways to make the inserts machine-identifiable. One would be some form of barcode; another would be to equip them with an RFID chip. Printing a barcode on a complex-shaped little piece of metal might not be easy, particularly when considering the reliability required, and writing it by means of a laser might prove more expensive than using a chip. Whatever they use, the longer they wait, the more the balance will tilt toward the chip solution, at least as far as cost is concerned.

So, this is an example of a true mass customization production. Inserts are produced automatically to an individual code and fitted automatically in customer-specified cylinder types.

Processing Glass

Next I will introduce a second example of smart manufacturing: the production of insulation glass units. This process is much more complex and requires integrating much more machinery than the master key example discussed earlier.

IGUs are assembled on a sealing line. For a double-glazed unit, two pieces of glass are put onto the beginning of the line, and for a triple-glazed unit, three pieces are used. The pieces can be put on manually or by an automatic feeder (I will come back to that).

First the pieces of glass run through a washing machine and are then inspected to see whether there are any defects. Then a spacer frame is applied to the rear piece of the unit either manually or by an automatic applicator. Some lines have an automatic turning station that will turn coated pieces to make sure the coat is in the right place, usually on the inside of the unit. After that the pieces run into the press, the front piece is picked up by the press and moved over the rear one, and then the pieces are pressed and glued onto the spacer frame. At the last station, the units are airtight sealed all around, along the spacer frame.

For standard rectangular IGUs, this is not rocket science. In fact, the sealing line can do it all automatically, as long as all the parts supplied for a unit are the same size. It becomes more complex, however, if sizes are not rectangular or if some of the pieces overlap. Then the sealing line needs a bit more information and has to be fed with sealing code.

Running all this does not yet make an IGU manufacturing business a smart factory. This comes into play only if you include the production of the glass pieces too and if you chain the two production lines, such that producing the IGUs from glass stock to final units requires little or no human intervention.

This is easier said than done. There are different types of glass used in an IGU, and some of them can be quite expensive. To minimize waste, all the pieces of the same glass type in a production batch are collected into layouts, and an optimal layout of those pieces on stock sheets is calculated.

Cutting and breaking that glass, however, will now not produce the pieces in the sequence needed at the sealing line. Traditionally this was dealt with by placing the pieces on trolleys and then sorting the trollies and bringing them in proper sequence to the sealing line. However, this required a lot of manual handling of the pieces, which not only is work intensive, but there is an added risk of injury and damage to the pieces.

The next step was to automatically sort the pieces into harp racks, which are trolleys that have multiple slots for single pieces. Sorting into these allows the factory workers to roll the trolleys to the sealing lines where the pieces are automatically pulled out and moved onto the line by an automatic feeder. The disadvantage of that approach is that the trolleys still have to be moved manually, and they can become quite heavy.

Thus, the tendency in recent years has become to place pieces into a big sorting array instead. These have hundreds of slots. Pieces are loaded into a shuttle, which then sorts them into the slots. Another shuttle later removes the pieces from the other side and brings them to the sealing line. These systems work well as long as all the operations are straightforward.

Now we have a long automated production line, where in many cases nobody has to touch a piece of glass until the final sealed unit is removed from the end of the sealing line. This saves a few jobs since the pieces are now moved completely automatically.

CHAPTER 3

Problems in Smart Factory Projects

"We only think when we are confronted with a problem."

—John Dewey

Why Projects Fail

There are a lot of risks involved when a new investment project starts. The better you understand those risks, the better you can mitigate them.

During the first and second industrial revolutions, most manufacturing companies were managed by engineers who understood the technologies and thus to some extent the risks they were taking. However, history teaches us that even back then there were projects that failed, some of them spectacularly.

To me that is no surprise. The bigger the project, the higher its complexity. Combined with a lack of knowledge, these are the highest risk factors. And modern automation projects are highly complex.

If you were ever involved in a project that failed, had a bad budget overrun, or did not fully deliver, then of course you try to be cautious next time. But what can you do?

Chapter 3 | Problems in Smart Factory Projects

In this chapter, I give you some understanding of the problems such a project faces. I will not give you advice on how to mitigate the risks. The right methods depend largely on what you want to do and where you sit in that project. But if you understand the problems, then you can figure out what you have to do. If you are the chief executive officer (CEO), then the main thing to do is to have the right people in the right places. I will talk about how to identify the right people later.

Increasing Complexity

In the olden days, all automatons were controlled by complex mechanical systems. In the 18th century there was already an automaton that could write, and you could change the text it was writing by changing the characters screwed on a wheel that held the "program." Remember those old punchcard-controlled organs that are sometimes displayed at fairs and exhibitions? Way more complex but based on a similar principle were the Jacquard weaving machines, which allowed weaving complex patterns.

These days, the mechanics have in that respect become simple. The mechanics are all controlled by electronics, which means by computer programs in some form or other. In the early days of that development, the programming part was fairly simple, but I remember a project where the machine was completely built and just waiting in the cellar for the electronics to arrive. The electronics department was so far behind schedule that the machine had to wait for more than a year!

Ten years ago, most automation projects were still reasonably simple. Programmable logic controllers (PLCs) were used and are still used where appropriate. Compared to large commercial software systems comprising millions of lines of code, PLC programs were simple and most of the time straightforward to program.

When in 2005 George Stepanek published the book *Software Project Secrets* [4], nobody outside the software development community seemed to notice, despite the subtitle *Why Software Projects Fail*. After some 60 pages of problem analysis, the book was mainly advocating agile software development methodologies, and that's why even in the software development community the book was mainly preaching to the converted.

In complex automation projects, we now see similar phenomena to those that Stepanek was describing in 2005 for software projects. And that's no surprise, considering that all the complexities of those projects now lie in the software embedded throughout and around those machines. This can start as far away as your client's office, where she electronically orders on a server in a private cloud or over the Internet directly. This is followed by order processing and ordering of external components and then by production planning and passing a production batch with all the relevant data to the manufacturing execution system and to the machines.

To understand better where the problems are, let's revisit the first chapter in Stepanek's book. Stepanek cites a report of the Standish Group that in the year 2000 only 28 percent of software projects succeeded outright. Twenty-three percent failed completely, and the remaining 49 percent were substantially late, over budget or lacking features, or often a combination of these.

In many automation projects, I have started observing similar issues, although not to that extent. After all, the biggest part of the cost of automation projects is still in the machinery, so if the software part goes over budget, then the software is still a minor part of the overall cost. But the other two issues start biting a bit more. After all, if the machinery is installed but not working, then that is a lot of money not working. And if some required features are simply not working in the end, then it means that some minor or major part of the plant is not working as efficiently as it should.

12 Reasons Why Automation Projects Are Different

Stepanek gave in his book 12 reasons why software development is different from, say, a building project. After all, 94 percent of customers of such projects say that their project was completed to their satisfaction. So, something must be significantly different. If similar effects are taking place in large automation projects as there are in software projects, it might be worth a closer look at some of those 12 reasons and see how they apply here.

Software Is Complex

What every programmer learns early in his training is the KISS principle: keep it simple, stupid. However, as anyone experiences quickly, you can try to make the software as simple as possible, but if you make it simpler than that, then the program will not perform as required.

Or as George Stepanek has put it, "Software is unique in that its most significant issue is complexity."

The size of a software project is typically counted in lines of code. If you write exactly one instruction per line, then that somehow makes sense, although I sometimes write multiple simple instructions on one line, and on the other hand some complex instructions can take multiple lines.

You could compare every instruction with one of those tiny moving parts in the 18th- and 19th-century automatons. Those things were complex and contained hundreds of parts. Probably the most complex of them all must have been Charles Babbage's mechanical computer, which unfortunately was never completed.

While all those systems were complex enough and contained hundreds if not a few thousand moving parts like wheels and levers, even a fairly small program runs easily into tens if not hundreds of thousands of lines of code. What we encounter now in our automation projects are hundreds or thousands of moving parts controlled by hundreds of thousands if not millions of lines of code!

Sometimes we succeed in finding the right abstractions that not only greatly simplify the code but make it work more efficiently at the same time. However, this is not for all to see and understand, or as Edsger Dijkstra wrote, "Simplicity is a great virtue but it requires hard work to achieve it and education to appreciate it. And to make matters worse: complexity sells better" [5]. If your software developer manages to find the right solution to make your system run elegantly, smoothly, and easily, then it's sometimes hard to understand how much abstraction and complexity the developer had to handle to come up with that easy-to-use solution. Even in the software development community this sometimes shines through. Recently I have seen several papers on software maintenance suggesting that highly abstract, elegant, and efficient solutions be replaced by more complex and less efficient ones so that the maintenance staff can understand them.

Software Is Abstract

You can store software on your hard drive or on a memory stick, but you can't physically touch it. It describes a set of behaviors that in isolation are easily understood but in combination are hard to comprehend. In particular, if hundreds or thousands of them come together, it's hard to understand all the possibilities.

When building a machine, you can design all its parts and draw an assembly plan that shows how it all comes together. In contrast, software encodes the behaviors that the users have asked for. You can't draw a plan of it first. The program code is the final specification. Once it is worked out, the product is finished.

Any design or blueprint of a piece of software must simplify the product to be comprehensible. But by doing so, it necessarily becomes incorrect because removing detail can mean not describing the part that later makes the program fail. The following point is an immediate consequence of complex and abstract software.

Requirements Are Incomplete

In his book Stepanek argues that managers and users are experts in their own roles. Some users have a good understanding of how it all works together, but they lack the ability of a professional software developer in dealing with abstraction and complexity. Even for someone with a good understanding of how it all works, it is difficult to consider all the alternative interaction flows and how different requirements relate to each other.

While this is all correct, at this point I would like to go one step further and state that while users and managers might understand the current processes, this view reflects only the old approach to IT, where we tried to replace paper-based solutions with computer-based ones. However, in modern development projects we replace outdated solutions with new ones, and in our automation projects we replace an old way of manufacturing with a new and ideally better one.

If you want to do that, you have an additional problem: users and managers are not requirements engineers or software project managers. Users have a lack of vision. It is their job to keep the operations running smoothly. They might know the bits and pieces of the existing system that are not well-designed and would like them to be improved. But often they have no idea of what they could have with modern technology instead.

So, if your production manager is given the job to modernize the factory, then he does what he has always done. He contacts machinery manufacturers, goes to exhibitions, and gets quotes for machines that he figures could improve his existing production processes. He might even evaluate new production technologies and get quotes for machinery that can do that. But when it comes to the information and communications technology (ICT) part of all that, he simply expects the software to resolve all the arising complexities.

You can try to involve the chief information officer (CIO) with that project as well. While that might help sometimes, more likely than not the CIO will come into this project merely to resist change. The CIO has his own problems to solve.

Maybe he tries to dig himself and the whole ICT department out of a big heap of legacy code that still essentially runs the business. If he can't upgrade and modernize, he is stuck with patchwork and will no longer be able to provide an adequate solution for the evolving factory.

All CIOs have problems with security. While the threats become more and more diversified, his budget is not growing adequately. This could mean he does not have the people to adequately deal with security risks. This in turn could mean that he will try to minimize the connectivity of his network to the outside world.

If the CIO resists change, then his message is that he does not have any programming resources to help with this project, that he does not want to integrate the machine network with the business network, or that he does not allow any wireless networks in the factory.

So much for the grand vision for the new project. What about the machinery manufacturers? When looking at my projects in the glass processing industry, it certainly seems to be the case that they had all the solutions. If you go to the GlassTec exhibition in Düsseldorf, Germany, or to Vitrum in Milan, Italy, some of the bigger manufacturers have rented essentially whole exhibition halls. In there they have built fully integrated systems of machinery, and it all seems to run smoothly. They can arrange visits to some other factories as well, somewhere far away from your place, so you can't spy on your competitors.

However, if you take a closer look, you might find that this all runs only in a much simpler context than what you need, that it does not integrate with your order processing or dispatch, that your best products cannot be produced, that some of your ideas cannot be implemented, or that they are very expensive to implement.

There is another problem I will discuss here. If the board decides on a new project, then this means there is money available to improve things. Quite often at the early stages of the project it is not really clear how much money you have access to because your board would like first to get a bit of an idea of what should be done or could be done and how much everything would cost.

If the department that will get the money has received little money for innovation purposes over a long period of time, then they know that this is the opportunity of their lifetimes. They have to come up with everything they could possibly ask for and make it all essential because after such a big investment, chances are small that they will get more money two or three years down the track. Thus, they will even ask for things they are not sure will be needed in the future.

I've seen this happen in pure software projects, where no machinery was attached anywhere. I was working with a friend at the time, using the new agile approach. We both worked from home and met once a week to integrate the new parts we had developed. We then discussed what each of us would do next, and we did a bit of programming together where we needed both competencies, first to make the integration work and then to extend the interface so each of us could continue with our part for another week.

Then we got a project for which I had done an estimate several years earlier. Unbeknown to me, the project had grown larger in the minds of the users and managers. I was told of some of the ideas, but they said that we would implement those ideas at a later stage. I tried to convince them to work in an agile manner, doing two-week iterations, and to have an early version that we could release, with the basic functionality only, and then to learn and grow from there. However, I did not have a chance in the world to get agreement on my idea, and since we were charging an hourly rate and there was no explicit talk about money anyway, we started with the job by trying to design everything first.

As the project grew and started to show functionality, the wish list started growing too. The users started to see the potential and asked for more and more functionality, and of course since there would be no more money after that, they asked for all the other stuff as well as what they saw they might need in the future.

For every additional function, we tried to give them some estimates. We always estimated hours since we could charge them by the hour, and it would be their job to do the numbers. I still don't know what happened exactly, but at some point the top management had had enough and stopped the project. It was in a stage where all the essential functionality was working, so it was released in that form.

But there was some other functionality in the project that we knew would help, largely with other functionality the business would need to modernize in the near future. This was not an easy part, but because I had written the old version of it, I knew exactly what had to be done, and I was almost finished with that job when the project was stopped.

Of course, we had to take the blame for the cost overrun and were never given any major project again by that company. Fifteen years later and with their third attempt, I'm still not sure they'll manage to replace my now over 30-year-old program. Some of the features they then thought that they might need in the future have quietly been removed (45 percent of the features implemented in software projects are never used, and 19 percent rarely, according to the Standish Groups Chaos Report 2002).

So much for this sad story. What strikes me is that in a recent automation project I started to see similar patterns. It wasn't that the project failed but that people tried to pack in all the bells and whistles they could think of. This started with the specification of the machinery, where they came up with a wild combination of standard components of the machinery manufacturer. It only later became apparent that not all of those components were designed to work together, and it took the machinery manufacturer a big effort on the PLC side to make it all work. When we started to analyze the result, we found more issues and had to point out that we still had to exclude some special cases because they could not be handled by the machinery.

What should have been done was a thorough analysis to calculate the productivity gain of these features compared to their cost.

Technology Changes Rapidly

Technology changes rapidly is the next point that Stepanek makes. However, in his book he talks only about ICT and even points out that technology changes much quicker than in other fields. After all, he tries to point out that software projects are different.

However, ICT is in almost every product or its design process somewhere, and ICT has accelerated the product change process in many other technologies as well. This means, of course, that other industries encounter now or will soon encounter similar problems to those encountered in ICT some 20 or 30 years ago.

What you can see now in ICT is a rapid movement to the cloud for many applications. There are many tools out there to build cloud applications and many ways to build them. With all the improvements and extensions to the Hypertext Markup Language (HTML) standard proposed or recently implemented, there are still a lot of extensions and improvements to the programming technologies to be expected as well. You'll find an in-depth discussion of what cloud solutions mean for manufacturing in the third part of the book.

Best Practices Are Not Mature

In other fields of technology, this point to me seems to be an immediate consequence of technology changing rapidly. What Stepanek tries to point out, though, is that even with programming, there is not one set of best practices that fits all languages and tools.

Of course, there should be documentation, and in the industry we all agree that all documentation should be in the code because any other documentation gets outdated quickly anyway. But after that, best practice really depends on the tools. As a simple example, if a particular programming language does not have a go-to statement, then you don't need a standard that tells you under what circumstances you do or do not allow its use.

When implementing a mass customization project, then this means programmers will have to use the newest technologies and tools, and it might well be that some of the manufacturing technologies are new as well. Thus, for those projects, best practices cannot be mature.

Technology Experience Is Incomplete

This is an immediate consequence of the last two points. If you try to use new technologies you think are mature enough to do the job, then that means in every new project you will use new technologies where you do not have enough expertise.

Think of it for a moment in terms of your new automation project. New technologies in the ICT part are a given. You will have to live with that. But outside of that, do you want to use new technologies to produce your products? Or do you want to produce new products that use or provide new technologies? Or all of that? Welcome to our world!

There is risk in that, and you know that. A good risk mitigation strategy could help a lot. In software development, the best risk mitigation strategy is agile development. That's what Stepanek's book is about. But there are other risks in ICT investment that need attention. You have seen some already, and I will discuss more later.

Technology Is a Vast Domain

In ICT it is certainly true that nobody is expert enough anymore to understand all the technologies used to make even a medium-sized project work successfully. You might use a database and a programming language to build the applications. The programmers understand how to interface with the database, and they know what types of queries and updates they run. Usually they understand a bit about the proper architecture of a database, but in big applications this is better left to specialist architects.

Knowing how to interface with a database does not mean you know how to run it, when and how to do a backup, how to set all the parameters so it runs optimally for the given applications, or how to analyze load and tweak the system to make it perform better.

And then there is the network. Neither the programmer nor the architect nor the database manager knows enough to run the network efficiently or knows how to analyze load and how to optimize.

I did not even mention the security issues. We all have to understand each other's problems so we can communicate more efficiently and cater for our partner's needs, but we need specialists in every field to make the thing run smoothly. And because our technology experience is incomplete this will not be the case the first time the new app runs under a full load.

Again, think of your new automation project. Isn't a lot of that true for your project too? If you just set out to do the old thing the old way with a new machine because the old one will break down soon, then of course this is not the case. If that is all you want, then you might just waste your time reading this book. But if you take the opportunity to improve a few things, then there are elements of this in your project. If one of the new technologies does not work for you, then the whole machinery you bought might just be a waste of money. Where that becomes obvious you will be very careful. You might want to do a pilot project to prove the technology.

Sometimes in ICT these risks are not as obvious. However, if you can see them, you usually do a pilot project as well, sometimes just to evaluate technologies and in other cases to figure out how the system will perform under a full load. However, at the early stages, it is sometimes quite difficult to know what *full load* really means.

The domain of the new technologies might not be as vast in manufacturing as in ICT. But the risks are there and need to be mitigated.

Software Development Is Research

This is, as Stepanek explains, the result of the previous two points. He states, "Software development isn't just a process of creating software; it's also a process of learning how to create the software that is best suited for its purpose."

Again, if your new automation project is about research, then you usually understand and do the research beforehand, before you invest all the millions into machinery that might in the end not do the intended job. What is different in software development, though, is that this holds for most software projects of reasonable size. And you cannot build a pilot project that does the full thing just in a small volume. The only way to save a little bit of money is by not investing into all the hardware before you have all the software.

Repetitive Work Is Automated

Well, this is in a sense a software special. What Stepanek tries to explain here is that software development is not like brick laying, where you know that it costs, say, $1 per brick, and if you have a wall of 10,000 bricks, then $10,000 is your price. In software, if I find that I have to do a similar job 10,000 times, then I write a program that does that for me.

This book is about automating repetitive work. We have done that in software development from the start. In mass customization, we do a similar thing, although we don't simply construct software code; we produce products.

In the olden days, automation meant producing thousands of identical products. This book in contrast is about producing thousands of variants of similar products. It's like a method or a procedure in software that produces a different outcome depending on the values of the parameters passed in. I will talk about machinery that can handle that, given the right data and software. And I will talk about software that generates the right data.

Construction Is Actually Design

This point is something unavoidable in software construction, but it's something you want to avoid when you build your new plant. In a building project, you can draw all the plans first and define all the work, and then you can estimate well the cost of the actual construction.

What you can do in a reasonably detailed software concept is try to thoroughly understand what the job at hand is and what all the details and exceptions are and then come up with some estimates for implementing them.

During implementation, however, it is unavoidable that you have to work out in detail what the user needs. And quite often this is when the user discovers additional modalities and exceptions that have to be handled there. (See the sections "Why Projects Fail" and "Increasing Complexity" in this chapter. If you try to figure out how the new tools work, then you find features that are not exactly doing what you need and others that suggest that it would be easier to do the job in some other way than planned.

Change Is Considered Easy

Often, if you sit with users to build the user interface, then they see how easy it is to create and modify that interface and even to implement some basic functionality. What they don't see is the work that follows afterward that makes sure that all the data entered is processed correctly, that all the data displayed is properly updated, and that all the error conditions are handled.

Often users come later with a little wish, and most of the time it's quite easy to do. That's why they then struggle to understand why another seemingly simple change would mean so much work. Unfortunately, it is not always easy to explain everything that is involved in that seemingly little change.

If their "little wish" implicates changes to the workflow, then you are sometimes able to explain what is affected. If that's not possible, because everybody knows that software companies make their money by overcharging for later extensions (and a lot of them really do!), these situations can potentially create tensions even in situations where there was reasonable trust established before.

Change Is Inevitable

Because construction is design, where we work out the details and where the customer usually figures out the details of his specification, there will be desire for changes and extensions. And because these changes are often quite easy, we simply do them.

Chapter 3 | Problems in Smart Factory Projects

But because changing software is considered easy, the other problem we are facing is that customers want us to change our software when they discover a problem. It can become a bit painful if we have to point out that the machinery ordered can't perform what they need. That's why we like to be involved early in the process. Unfortunately, that's not always what happens. If we are involved early, we can ask all the questions we have to ask, and we can make sure all the requirements we see are written into the contract with the machinery manufacturers.

Big problems can arise when a change means a bit more of an effort. I always try to make the customer aware of this possibility. If they are aware of it, then they have a budget for that. However, in automation projects this can mean changes on the side of the machinery manufacturer as well. If they were hard-pressed to deliver at the lowest possible price, then they will behave like software houses too: they will charge as much as they can get for every little change.

Usually, what we need in those situations is just a bit of an extension to the software interface to those machines. On the machine side, it usually needs changes to the PLC program; changes to the machines themselves are rare. To avoid this situation, you usually want to try to get an interface that is as flexible as possible, meaning you can provide the machinery with every little detail that you want it to perform so you have detailed control over the system. If the machine manufacturer already has a standard interface, then during the quoting phase you can usually point out only its weaknesses and what you would need to add to make it work optimally.

Conclusions

To sum up, you have seen that most of the 12 reasons that make software projects different from, say, building projects, are now valid for sophisticated automation projects (see Figure 3-1). If you consider that software is a significant part of such projects, then you see that in fact all 12 points can become aspects of your project.

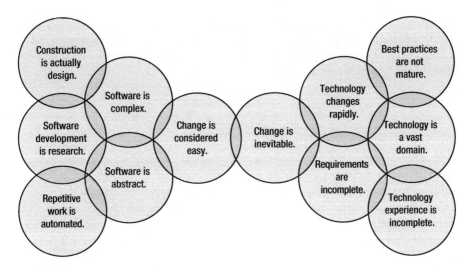

Figure 3-1. Twelve reasons why automation projects fail

Stepanek then makes the point that in many projects other than software a few of those points might be true. They're not always the same, but typically two or three are. And of course not all of those aspects are valid in a software project, but usually a majority of them are, typically more than seven or eight.

Stepanek then goes on to state that if in your project more than four of these aspects are true, then you should no longer treat it as a project but rather as a process. In the software development community, this gave rise to the concept of agile development. But that can't always be translated easily into the development of large automation projects.

However, if you go back and read up on the reasons that were given over the last 15 years on why agile would not work in a particular environment, then you'll see it was almost always because some people in the business could not get their heads around that new paradigm.

On the other hand, if you go to the list of reasons covered in this chapter and think about those that are valid for your project, then you will understand that you have to make a start somewhere, even if you don't have all the right answers yet. And that's when you have to start a process and become agile.

The idea is that you come up with a goal and a road map on how you want to get there. Try to pick the low-hanging fruit first and do those things in the initial steps that give you the shortest return on your investments. This might not always be possible; sometimes you will have to build a critical bit of infrastructure first, for example a factory hall. But even in those cases, try to build only as much as you reasonably can and will need for the next few phases. I will discuss some of those aspects a bit more in depth throughout the book.

CHAPTER

4

Future Challenges

"One never notices what has been done; one can only see what has to be done."

—Marie Curie

Regardless of whether you call it a *smart factory*, *mass customization*, or *Industry 4.0*, there are challenges that need to be addressed. The executive summary of Germany's Industry 4.0 report [1] nicely sums up the most important ones. I will discuss those challenges briefly here. Later in this book I will have a bit more to say about some of them.

Standardization

Obviously there is a lot to do in this field. Standardization has to happen on many levels. First, it needs to happen on the network level, where standard protocols and interfaces are needed across all types of network technologies. In the long run, it should no longer matter what technology is used; the interface for a programmer should always be the same. However, this is easier said than done, in particular if technology continues to evolve as fast as it is evolving right now. There are standardization attempts under way in many fields, but the problem is that there are too many of these attempts in most fields. For the Internet of Things (IoT) alone I am aware of half a dozen or so standardization initiatives, and the field is still evolving fast.

Of course, this is not all that has to be standardized. At the top level, there should be a way of standardizing the interface between all the participating companies in a network. Again, this is easier said than done. Obviously, in a technical sense there are so many different products that there can't be a global standard that allows communication between all customers and all manufacturers.

In the glass and window business, there are about as many interfaces as there are window manufacturing software providers, and the supplying glass manufacturers have no choice than to implement them all. What makes things worse is that most of them have their roots in the olden DOS times and are limited, in that they don't allow specifying all that can be produced. In those cases, the good old fax machine has to be used.

And then there is everything in between. I don't want to blame anyone here; in all these cases, interfaces have grown, and additional functionality has been added, but still there are things that can't be handled. For example, machine interfaces are not standardized, not even for similar types of machines, such as glass cutters or routers. The router world is a little bit better in that most manufacturers at least use the same style of code (called *G-code*), but the details are still different, and you have to write machine-specific code for every machine.

What I'm trying to say here is that you have to be pragmatic. Interfaces evolve with technology, and although standards try to go with that, they struggle to keep up. So, whatever technology you want to use, you will have to use what you can get at that point in time. On rare occasions in the past I found this strategy lead to a really bad choice and had to change later, but in the vast majority of cases, once implemented, I can live with it.

Complexity

The Industry 4.0 report stresses the importance of modeling for companies, in particular small to medium enterprises (SMEs), to be able to handle the complexity of modern factory automation. I fully agree with this view. But I have to say too that SMEs in particular will struggle to run highly complex factories efficiently.

Let's say, for example, you are producing beds and want to let the customers configure their dream bed online. If you offer, for example, five bed sizes, two styles, four types of material each, six types of bed bases, and four types of bed heads, then that gives you 960 different beds. You could split it up, but you would still end up with 160 different heads and 240 bases. You could try to commission this from stock, but if you want to have efficient batch sizes, then this requires a lot of stock, and beds are neither small nor cheap, so they are expensive to stock.

In addition, producing it all to order might become a bit of a pain because you have so much variation over your production line. At least you can group by material and then style and size. However, there might be some rarely used materials or bed sizes, so it might help to have some of those on stock in order to produce them in small batches, maybe once a month. Modeling and simulation can help you a lot if you try to figure out what the right mix should be.

Suppose you have an automation project that has the potential to produce almost all variants of all your products automatically, but running the plant seems a bit complicated. In that case, I strongly recommend doing a simulation first. Your system integrator should be able to do this. Try to run as many load scenarios and use cases as you can to get a good picture of the overall system performance and the weaknesses of the system.

In rare cases, you might find that it's all fine and the complicated plant should work without problems. In many cases, you will find that you could easily improve the system to make things work better—sometimes on the machinery side, sometimes on the software side, and usually on both. And in a lot of cases you will find that it's simply not worth having all those bells and whistles that make it work in every way you could imagine.

Try to simplify it instead. Build the system so that it still does the bulk of the work, and accept that this will increase the percentage of jobs that need a more traditional way of manufacturing. Even for those traditional cases, you are still able to give your workers on the floor a lot more support when using the newest information and communications technology (ICT). I will discuss these aspects later in this book as well.

Broadband Infrastructure

Most certainly the countries that manage to develop a good broadband infrastructure will have a higher success rate with those new technologies. It might not be important to deliver a 1 GBps connection to every household, but we have to make sure every business, even every small business, gets the connection rate it needs at a price it can afford.

You have to be aware as well that with the Internet of Things (IoT), the number of wireless connections needed by all those devices will explode. The best way to cope with that would be to link back into the wired network wherever possible, allowing you to reuse frequency bands for wireless relatively close by again.

Other than that, you will have to make sure the broadband infrastructure within your company is capable of handling the internal data flow. If you need a lot of wireless capacity on the factory floor or in the office, then LiFi might be the way to go. (I will discuss that later in the book; it's an emerging technology that uses LED down lights to carry a signal.) Most likely you will have more LED lights than computers in your office, so there would be a capacity of 1 GBps download speed for every computer and still plenty left for mobile devices.

Safety and Security

I'm neither a safety nor a security expert, but I can say for sure that there will be challenges. However, with recent developments of robots that have sensors and are aware of their environment, it is becoming safe to have robots working side by side with workers. There might still be a way to go, but I expect that there will be a lot achieved in the near future.

As far as security is concerned, this is a challenge that we all face already. It will no longer be acceptable to try to shield your network from the outside world. If your network and security experts still try that, then I'm sure they have already failed. In the future, this will become worse. They will never get there. So, instead of letting them try to go down that path, tell them to learn how to mitigate the risks they are confronted with.

Work Organization and Design

This is an important point. In a smart factory, workers have to be smart too. In his speech [6], Prof. Goran Roos pointed out that what is currently the highest qualification in many factories, a bachelor's degree in engineering, will in the future be the lowest degree in a factory.

That immediately reminded me of a speech I attended maybe 30 years ago. Back then I was working in a factory that had several departments where automata churned out thousands of identical pieces for electrical devices every day. The company built control panels as well that went into automata that churned out thousands of parts for car manufacturers every day. That speech was about automation, and the speaker pointed out that the European manufacturers in that field tended to employ untrained people who were not able to sense that something was wrong with a machine before it stopped, usually with a bang. In contrast, Japanese manufacturers employed engineers in the same places.

I think it depends on the actual situation when deciding on who handles a particular process. But it certainly helps if the worker in charge understands what the machine is doing. In a smart factory, this can mean the worker has to understand a lot about the context in which he operates.

Training and Professional Development

I will discuss this point shortly in a bit more depth. Changing from a paper-based to a computer-based manufacturing world can be quite challenging for many of your workers. In many cases, the workers all understood their jobs quite well, and a quick look at the paperwork was sufficient for them to know what to do next. They knew where to find the information relevant to them and what to do with it.

Now everything is new. There is an application running on the screen with a menu, and they don't know where to go. These guys are not used to going back to school to learn new stuff. They learn on the job, and if there is so much new stuff to learn, this will take time. So, please be patient. Don't show them a hundred times, though. Let them do it themselves, observe for a while, but intervene only if you must.

Most likely the guys on the top floor will have to learn a few new things too. This should be a little bit easier. They are usually more used to change. However, what has changed now is how the factory operates. Production planning and scheduling will no longer be the same.

Some of the added complexity has to be handled on the top floor as well, and often that is not easy. If you did a simulation and already know how the system should operate, then this will help. The simulation model could even help you to train those people in advance.

Regulatory Framework

I'm not an expert in this field, but what was stressed in the report is certainly important and can be understood by all of us. In an environment where anything is produced automatically and individually, it can become difficult to see and monitor whether a job entered into the system leads to a legal product, which fulfils all security and safety requirements and is not under some sort of legal control, such as weapons.

In many cases, it will be a challenge to make sure nothing goes wrong here. Instead of trying to regulate with detailed laws (which will always be late), there is a call for a legal framework and technical measures to help stop something going wrong.

Resource Efficiency

The argument here is that an intelligent machine is more resource efficient. For example, if a welding robot knows its operational hours, then it can shut itself down outside those hours and start up in time to be ready for the next operation.

I am of the opinion that being resource efficient is a constant task. If a little device that can shut the welding robot down outside operational hours costs more than what you can save in a reasonable time, then you won't make the investment. But it certainly helps to look at efficiency issues again as prices change, be it the price of an energy source or a material you use. Prices can change dramatically, or they can change a large extent over a longer period of time. Either way could alter the investment equation.

PART

II

New and Emerging Technologies

"Somewhere, something incredible is waiting to be known."

—Carl Sagan

Part II of this book looks at a suite of new technologies that will influence the way we will be manufacturing in the near future.

First we look at new developments in information and communications technology (ICT) that are already changing the administrative side of things. Then we look at new and emerging technologies that are about to change not only how we communicate and control machinery but also how we communicate with people on the floor.

Next we look at new manufacturing technologies, most of which have become available or affordable because of recent ICT advances, and at a few others that will extend the things we can produce and the way we can produce current and new products.

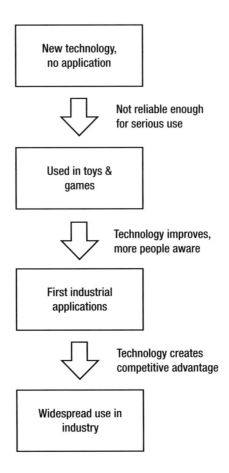

Many of those technologies have been developed over a long period of time and were not taken seriously because they did not produce anything serious. However, innovation always starts with someone playing around with some new technology that doesn't seem to work well just yet. After a while, someone tries to make money from it, and because it's not reliable enough and because nobody who knows the technology knows about "real" applications, the first commercial products are usually toys and games.

Next some people with kids and "real" problems see an opportunity in those toys, and the first industrial uses emerge. Examples of new technologies with which we are at this stage at the moment are 3D design on the Web and 3D printing, especially additive manufacturing.

It is important to point out that these two technologies go together very well! In a year or two there will be applications where you can design your 3D gadget online and either print a model of it (if the real thing is too big or too small) or send the data to your favorite additive manufacturing supplier to produce that part in the required material.

Some of the technologies discussed here might still be at the toy stage or just emerging from it. If you observe such a development in a technology that might be relevant for your industry, try to get a head start, be an early adopter, find meaningful ways to use the technology, or try to improve it so it becomes useful to you.

Of course, I do not know all the new technologies that researchers are working on, but you should be aware of those in your field. All you have to do is to have an open mind, observe progress well, and try to get in when the time is right.

I'll start with an example that everyone is aware of; it might not have much to do with manufacturing but can illustrate well where things can take you if you keep an open mind.

CHAPTER 5

New ICT Technologies

"To err is human, but to really foul things up requires a computer."

—Anonymous

Smartphones and Tablets

Smartphones and tablets (and anything in between, for example palm tops or the iWatch) have been around for a while now. What is not so common is their use on the factory floor. However, this is emerging, and just recently I saw a demonstration of a warehousing application where smartphones were used to commission goods.

The people doing the commissioning could use their smartphones to see the warehouse layout, their own position, the positions of all their co-workers in the warehouse, and the position of the next item they have to pick.

When turning the smartphone to horizontal, they could see a picture and the article number of the next item they have to commission. One of the smart features of the solution presented was that it was producing statistics on the frequency of articles commissioned and then using that information to reorganize where the goods are stored to minimize the average distance that the workers had to walk.

One method of localizing smartphones in a warehouse is the magnetic fingerprinting method, as described, for example, in [8]. This method allows localization within a room with a precision of about 3 to 10 feet (1 to 3 meters). For it to work, however, the magnetic density pattern of the room cannot

Chapter 5 | New ICT Technologies

significantly change, an issue that might arise in many factory floor settings. I will discuss more options for indoor localization later in this chapter.

I can think of many uses for smartphones on the factory floor. One of the most important to me is its potential use as a barcode scanner. To be able to pack in a bit more information, my company inmatic started using quick response (QR) codes, but any 2D barcode would do in most cases. It is easy enough to write an app that can read that code and then report back or ask for further information over the smartphone's Wi-Fi connection.

This can be used to report progress. For every process on a worksheet, we have not only its due date and expected processing time but a QR code as well. If the QR code is scanned, the system assumes the job to be done by default, but you can override that and, for example, report a problem.

QR codes can be printed on labels as well and stuck to your pieces. In the cabinet-making example, you could encode the piece's description into the QR code. On the installation site, joiners could run a tablet that shows the whole kitchen, and when scanning a label with a smartphone, the tablet highlights where that part goes. To make the communication overhead a bit smaller, the smartphone could hook up to the tablet via Bluetooth so they would communicate directly without involving the server.

Small Computing Devices

The Raspberry Pi is a small computer the size of a credit card but a bit thicker. You plug in your TV and a keyboard and are ready to go. It is intended to show kids all over the world how computers work and to learn how to program. It runs on Linux and was initially sold for $35. The Arduino is another example for that type of device. Prices might have increased a little since then, but it's still cheap. The RFduino is an Arduino shrunk to the size of a fingernail with a Wi-Fi connection to the outside world.

Just recently, a new-and-improved version has been released, and there are dozens if not hundreds of add-ons available as well. If you Google *Raspberry Pi applications* or *Arduino*, you can find dozens of uses, such as controlling an unmanned aerial vehicle, automating your home environment, and creating robotics. IBM has even built a production-line monitoring system for a proof-of-concept demonstration. The Arduino is used by physicists to help with evaluating the mass of data coming out of the Large Hadron Collider and by a community of concerned citizens who built Geiger counter devices to monitor radioactivity around Fukushima.

It is these types of applications where I can see the use of such systems. In environments where you need high reliability or where there are safety issues, I would not try this. But if you can live with the fact that the system might go down on rare occasions, then this might really give you the chance to implement a simple and cheap solution to solve some of your manufacturing problems. After all, having some Raspberry Pis or Arduinos as spares in stock will not cost a lot!

3D Printing

To me there is a difference between 3D printing and additive manufacturing, although I will not try to draw a proper separation line between the technologies. Additive manufacturing is about producing parts for final products or for parts that directly help with production such as mandrels or one-to-one models that are then used to build the dies for die casting. You can use 3D printing in contrast to build models of something much bigger or smaller to help visualize, study, and understand it or, for example, to print a one-to-one model part to study how the real part can be fitted.

I will discuss additive manufacturing later in this chapter. Using 3D printing technology in a manufacturing context is not that simple. It is a tool for engineers. Its use requires a good understanding of 3D design.

To some extent, this will always be the case. However, it can be expected that easier-to-use software tools will come onto the market soon. This does not mean that this will automatically lead to good designs. But in the hands of an experienced designer, such a tool can become quite interesting. Even if he designs objects that are way too big for the average little 3D printer in the office, if he can use it to print a scaled-down version of what he is designing, then that will certainly help the discussion with the client.

The same holds for an architect. There is no longer a need to manually build models from wood and cardboard if it can just be printed instead. And this could be applied to not only the hotel from the outside but also the lobby if the customer wants to see that or a room, for example.

All these are easy-to-understand examples because someone is designing the real thing with a good understanding of 3D design. Extracting a part of that and scaling it down so a model can be printed is not that hard to program. The hard thing is the design part itself, which requires some understanding and design know-how. Some tools in this area have the necessary extensions already, for example AutoCAD, SolidWorks, or Photoshop.

Some of the people using such software packets are already using 3D printers quite extensively. Their major use in the manufacturing industry is for building models and prototypes. For example, if you want to build a part that goes into a tight spot, then it is often quite difficult to work out on the drawing board or in design software whether or not it is possible to install that part. However, printing that part and then trying to install it is easy. You might not be able to use that part because it is the wrong material, but simply figuring out whether it can be installed can help a lot.

As I said, the lines between 3D printing and additive manufacturing are blurred because you might be able to print some parts in the office on your 3D printer, which then can be used in a final industrial product. If a plastic part can do the trick and if you need only a few of them every time, give it a try!

There is news about 3D printing almost every day now. Just as I'm writing this book, there was an article that the U.S. information technology research group Gartner predicts some 108,000 printers will be sold in 2014, more than double that in 2015, and predicts 2.3 million will be sold in 2018. On the same day, the premier of Victoria, Australia, announced that every school in Victoria will get a 3D printer if his government is reelected. If that policy is adopted by many more governments in the first world, then the Gartner prediction might well be exceeded by a large amount.

Visible Light Communication and Positioning

As indicated earlier with the Internet of things, not all communication of all the machines and parts moving through your factory will have to communicate through the same medium. For safety reasons, some of the communication on the machines might still run over cables, radio-frequency identification (RFID) chips have their own standards, other things might use Bluetooth, and more communication technologies and protocols are developed all over the world.

VLC is one of those technologies that will help with other means of communication, should Wi-Fi get overcrowded in your factory. Combined with mobile devices such as smartphones or tablets or other mobile things in your factory, VLC can be extended to Visible Light Positioning (VLP), which gives many applications a big potential use in many factories. So, let's take a closer look at what this is.

Light has been used in many communication applications for a long time. Optical glass fibers typically use the near infrared spectrum; almost the whole Internet runs on that. In a newer development, light in the visible spectrum is used over plastic fibers as well. The nice thing about that type of connection is that you immediately see whether you have "contact."

Light-emitting diodes (LEDs) working in the near infrared spectrum have found widespread use too. You use them every day with your TV remote.

Since LEDs started to replace incandescent lights in recent years, we can now use these LEDs as signal transmitters as well. For this you don't need a specialist LED light; any commercially available LED will do. In fact, these LEDs can be used to carry a high data rate signal of around 1GBps. Please be aware that if you Google *VLC* or *LiFi*, you will find a lot of information from 2012 and earlier, when people were still trying to develop specific LEDs for that purpose and achieved lower transmission rates than what could be achieved only a year or two later over standard LEDs.

If traffic lights are changed over to LEDs, then they can start talking to each other as well. But any other LED light source could be used, such as a TV monitor or the screen of your smartphone.

Any photo diode would do as a receiver. However, these are not as widespread. There are smartphones available that measure ambient light to control the lighting of its screen, and tests were made using those. Cameras contain millions of photodiodes. However, if you want to use a camera or your smartphone, then this means high power consumption and thus short battery life.

The advantages of LiFi are as follows:

- High acceptance because people are already used to LED light and not afraid
- No flickering because of communication
- Easy-to-detect faults
- Works in lowly dimmed light

However, these are the disadvantages:

- There is only line-of-sight communication.
- Power drops off with the fourth power of the distance.
- It does not work in the dark (in other words, with the light off), but dimming is possible.
- Ambient light can interfere.

Another problem with Li-Fi is the return channel. Most of the time a Wi-Fi connection is used for that. If the downstream load is much bigger than the upstream, then this can still extend the capacity of your network by up to an order of 10.

In an industrial environment, the combination of VLC with VLP is of particular interest. One of the simplest examples of this is the search for resources in a hospital. Wheelchairs and other mobile resources tend to get "lost," such as when a nurse moves a patient back to his bed in a wheelchair and then gets an emergency call. Once the emergency is over, nobody remembers the wheelchair left behind, so nobody brings it back. Even the nurse most likely does not remember once the emergency is over.

In that situation, locating the wheelchair using VLC is particularly easy. If the lights in every room transmit their room ID only about once or twice a minute and if the wheelchair has a receiver, then the wheelchair knows long before it was even unloaded where it is. It can then just transmit via Wi-Fi its position, which will be recorded on the server.

In a large factory hall with dozens of lights, this becomes a bit more complicated, particularly if the position has to be more precise. As with GPS, three light sources are enough to give the position of a receiver. Using the same technology would mean, though, that we need to have clocks on each light source that are synchronized to one-millionth of a second. If that is possible for the GPS satellites, then of course it can be done for LED light sources too.

However, we are after a cheap solution, and synchronized time is not cheap. More promising in that respect is maybe the detection of the angles to the light sources, in other words, the angle of arrival at the receiver. This would mean a slightly more complex sensor on the receiver but would still yield a cheaper solution. At the time of writing, research is still being done in this area; see, for example, [9].

To my knowledge, one pilot application in this area is already underway in a shopping center. The shopping carts are equipped with sensors that link to a customer's smartphone via Bluetooth. When the customer walks the aisles, the cart always knows where it is. When reporting this, then the system can inform the customer of what's around him, the specials, his favorite products, and so on. The customer can download an app where he can enter his shopping list as well, and when an item on his list comes up, the system will tell him.

Now let's take a look at a proper production environment. One problem in many industries is finding work in progress when the next production step comes up. Or in the case of the glass customers, if a supplier delivers glass, for example, the toughening glass company to the insulation glass unit (IGU) manufacturer, then it would be helpful if the cart just reports to the system on arrival. If the supplier reported the delivery, then the system knows what is on that cart too, or if the cart is really smart, then it knows itself. The cart can then always report if its position has changed. If the system keeps track of the pieces on the carts, then it can easily direct the user to the right cart when she searches for a piece.

Because VLC and VLP work on a line-of-sight basis only, you can't use it for identifying the pieces themselves. There, a good old RFID might do a better job. If each cart has a receiver covering its area, then the pieces will check in and out with it automatically, and the cart can report to the system what it is carrying.

One last example for a neat solution is the use of a person's position via a receiver on a headset in a museum or a gallery. Because the system always knows the position of the user, it can then explain to him the exhibit he is looking at.

Case Study: Self-Driving Cars

Several announcements have been made recently regarding self-driving cars. Some are driving around already. Google's autonomous cars have driven more than 700000 miles (1.1 million kilometers) and have improved so much that they will soon be much better than us even in the urban jungle. Volvo and Mercedes have started similar projects. There is a lot of legal work to be done until self-driving cars can go into mass production and onto our roads worldwide. However, the process has already started, with several states in the United States having legalized driverless cars already and the United Kingdom following in 2015.

Where there is no new legislation needed, for example in mining operations, several projects and trials are already underway. Several mines in Australia are already operating with trucks without a driver. Maybe they will drive only to their last loading point and the operator of the digger has to direct them closer to his new place of work, but they will then find their way back to the unloading point, unload, and come back. It was estimated recently that by the end of 2014 half of all ore will be moved by driverless trucks out of Australia's mines.

While we wait for legislation to allow self-driving cars on public roads, car makers are already improving the system. Adaptive cruise control lets you follow the car in front of you at a safe distance. Essentially, cars drive at the set speed unless they get too close to the vehicle in front. This is a feature that drivers trust. However, combining this with active steering, meaning the car follows the one in front even into the overtaking lane, is much less trusted by drivers.

Systems that detect whether you depart from your lane are already in use. Cars that can park themselves are available too. Now there are systems under development that can find an empty parking lot and park the car there. New swerve-assist algorithms are using power steering and brakes to help you stabilize the car when you suddenly have to drive around an obstacle.

Nokia sold their mobile phone business and started investing heavily in technology where cars communicate with each other (this is called *vehicle-to-vehicle communication* [V2V]) and, for example, work out the best sequence in dense traffic. U.S. legislators are already working out rules for that type of interaction among vehicles while car manufacturers are working on communication standards.

If car makers build in more and more functionality that supports the driver but does not replace him altogether, then that gives all of us a chance to come to the conclusion over time that we can trust a self-driving car.

One other feature under development is trying to figure out whether you are drowsy. Interestingly, this can be done by just observing the movements of the steering wheel. Another system tracks eye movements, and there are more. In combination, they can quite reliably tell if you are drowsy. Can you imagine the system detecting that you are drowsy and then suggesting that you have a nap while it drives for you?

So, with all that under development, what do you think will happen in the near future? Self-driving cars will be on the roads as soon as they become available, and there are many operations where they can save money as well. Initially there might be a requirement for a human still sitting behind the steering wheel. But once that is no longer the case, taxis will be driverless. You will have to enter your address and pay up front. The taxi then brings you there and at the end charges your credit card the exact fare and prints your receipt. Neither taxi drivers nor passengers need to fear for their lives anymore when driving or taking a taxi late at night. Meanwhile, taxi companies start evaluating processes and algorithms that allow them to optimally combine parcel and passenger transport.

Public transport can make use of small driverless cars too. Imagine simply telling your smartphone where you need to go tomorrow and the time when you have to be there. The system will work out when you have to leave home and tell you that a car will be waiting for you at that time in front of your house. That will then drive you to the next train station where a train is waiting to bring you to a station near your target. There a car will pick you up again and bring you to your target.

According to a study from Intel that surveyed 12,000 people, about 44 percent of Americans would like to live in a city where driverless cars would operate on their own and connect public transport [7].

Let's spin this a bit further. When cars can be driverless, you can run a courier business without human couriers. Vehicles will find the closest parking lot if they can't park on the premises. They will advise the control center when they are near the target. This will prompt the control center to automatically call the customer, tell her where the car is, and ask her to pick up her parcel. The use of a camera in the car prevents customers from removing anything that's not theirs.

As soon as trucks get driverless, even manufacturers of bigger things can run their distribution without human drivers. This might in some cases be a bit more demanding since unloading becomes the full responsibility of the customer, and you would need to find ways to make sure that it is handled properly.

There will be something in that for us software developers as well, of course. Be it for couriers or for manufacturers, most likely they already have software that calculates their distribution tours. What you need next is software that is able to load that tour onto the cars' or trucks' navigation systems.

Mass Customization

Once trucks can be driven without drivers, police no longer need to make sure truckers don't drive overtime. Automatic driving systems don't get tired. Think about this: what will that mean to your distribution system?

Do you see the pattern? Google started playing around with a bit of new technology some two or three years ago, developing something that has nothing to do with your manufacturing business. Combined with a little bit of lateral thinking, we end up now with ideas that might revolutionize your business in the foreseeable future.

CHAPTER 6

Alternative Factory Floor Interface Devices

"The radical invents the views. When he has worn them out, the conservative adopts them."

—Mark Twain

If people have to wear heavy gloves and still need to give input to any machine more complex than a pedal to press, then emerging technologies can help. Some of these technologies have been around for a while; they are just not used on the factory floor. An example is Microsoft's Kinect, which allows you to give commands by waving your hands.

Gesture Recognition

One example for new technologies ready to be used on the factory floor is gesture recognition. The Kinect is an example of such a new technology, starting as a game and then finding its way into productive use.

Chapter 6 | Alternative Factory Floor Interface Devices

In its broadest interpretation, gesture recognition is trying to understand human gestures using mathematical algorithms. Gestures can be facial or bodily expressions or gestures made with your hands. In a factory context, emotion recognition from your facial expression is probably not the way to go. However, and particularly if workers have to use thick gloves and can't use a keyboard or a mouse, gestures made with the hands or arms can help to build a more complex interface than just a pedal, where at best you can have a forward/backward moving command.

In glass applications, in most places with a higher degree of automation people have moved away from wearing thick leather gloves. Even in places where occasionally a piece has to be turned or broken out manually, people use only thin gloves these days because they no longer have to move the pieces around, holding them at the sharp edges. If the pieces have to be moved, they hover on a cushion of air on a table and only have to be placed at a specific position for the system to then pick them up and move them on. With those white gloves, people are able to enter something at the keyboard or use the mouse.

One place where a gesture interface can help is at the end of the production line. There the final products are removed and placed on delivery racks. In those situations, you can display the rack on which you want to place the piece and where on the rack it should be placed.

In some instances, you need some feedback from the user, though. For example, you need to know whether in the end there is a problem with the piece or whether it is in some way damaged. If it were that simple, you could simply have a pedal or a button for "damaged," and then you would know the piece or unit had to be produced again and would not be on the delivery rack.

But as usual, the situation is a bit more complex than that. Manufacturers would like to know what type of damage they have in order to be able to improve their processes. This means users have to report the type of damage. What you need are a few simple types of waves with the hand or arm. For example, waving left-right-left could mean damage and will open the damage dialog with a list of damage types. Waving down could mean the active line will change one down and up the other way, and there could be another gesture, maybe using both hands, to pick a selection.

Stereo cameras, structured light, and other depth-aware technologies such basic gestures are easy to interpret. The Kinect is such a technology, and for quite some time now Microsoft has published the software development kit for it, so setting up such a system is easy.

If you want to tie the gestures to a specific user, you could use an input device as well, for example one tied to the user as an armband similar to the input device used by Nintendo's Wii. Make sure, though, that the system stops interpreting gestures if the user is simply going about his normal business.

Brainwriter

The Brainwriter of Impossible Labs is a device that works similarly to eye-tracking devices but is replacing the eye-tracking part with electroencephalography (EEG) simply read with a headband. This means the interface part of the eye-tracking device has to change, but all the rest can stay the same. In this way, people can control the cursor with their brain waves.

Although aimed at people with disabilities, this might be suited for people "disabled" by holding parts or wearing thick gloves.

Google Glass

People are divided about Google Glass. There are those who are using it, and there are those who don't like to be recorded without noticing. Google has already released a code of conduct to prevent explorers from becoming "glassholes" and to become better stewards of the technology instead. Another problem for Google is that once Google Glass lost its novelty, people lost interest in it.

However, adopting Google Glass for the workplace is nothing new and much less controversial. Surgeons use it in operating theaters, athletes use it in games, and there are many other applications in various industries. With its Glass for Work Explorer program, Google tried to regenerate interest in the technology, and rightly so.

On this or a similar technology, it would be easy to build a software interface for people on the factory floor. Other applications would, for example, be maintenance staff members who could report to their manager if they encountered some unusual issue, directly giving the manager the same view of the problem as the worker on the floor and still having an interface to talk to the manager via the Glass screen and mic.

Integrating Glass into your software system might still be a bit of a problem, but solutions could be available by the time this book goes to print.

There are other solutions on the market already. For example, Epson's Moverio is an augmented virtual reality overlay that occupies the whole field of view. This could be used to show you what is behind the panels you are looking at so you have to remove only the ones that are important now. This will solve different problems in the field from those that Google Glass does, so the two are not direct competitors.

CHAPTER 7

New Manufacturing Technologies

"As a final incentive before giving up a difficult task, try to imagine it successfully achieved by someone you violently dislike."

—K. Zenios

Nobody is an expert in all things manufacturing. So, in this chapter, I can only point out a few things I see coming. I'm aware of many of these new manufacturing technologies because they are directly or indirectly connected to new computer and software technologies.

Often it is the combination of two or three new technologies and several existing ones that lead to new products. I will point out some of the key enabling technologies here. Of course, information and communications technology (ICT) is one of the key enabling technologies in this context, and that is what this book in a manufacturing context is mainly about. But there are several other new manufacturing technologies available or on the horizon, such as nanotechnology, nano-electronics, photonics, additive manufacturing, digital manufacturing, advanced materials, advanced composites, and industrial biotechnology.

Chapter 7 | New Manufacturing Technologies

Some of the new technologies described here can be used to produce some of your products or parts in a new way (for example additive manufacturing) and become enablers of mass customization in that way. Others, for example assistive automation, could help automate your production of mass-customized products. Still others could be used both ways. For example, the Internet of Things could help you make your products smart, it could help you to make your machines smart, or it could do both.

The Internet of Things

I talked about the Internet of Things (IoT) briefly before, so here you will take a closer look at this concept and how it could affect your manufacturing business. Before I go into it, I will mention that the IoT together with all the other systems that in some way have access to the Internet (computers, laptops, tablets, palmtops and smartphones) have collectively been given the name Internet of Everything (IoE) or the Internet of Anything (IoA).

More and more widgets and gadgets have an ability to communicate. This has been a long time coming because radio-frequency identification (RFID) has been around for a long time. Basic RFID tags are simple and cheap and accept only a basic ID, but there are other, more expensive ones with a bit of intelligence used for access control, small payments, and other functions. They work by getting their energy from a reader over an electric field. They then start up, identify themselves, and exchange their information wirelessly with the reader.

The technology has come a long way, and there is now much more on the market than just RFID. In particular, there are many more ways to communicate and more ways to energize. At the March 2013 TSensors Summit, it was stated that the number of consumer sensors has grown from 10 million in 2007 (with the introduction of the iPhone) to 10 billion in 2013. And at their Stanford summit in October 2013, the organizers estimated that the total number of intelligent sensors will reach 1 trillion by 2022 [10].

Your smartphone is a good example of a (rather complex) multisensor system. It most likely contains a camera or two, a microphone, a touch screen, a gravity sensor, a GPS system, and maybe a few more. It can communicate over Bluetooth and Wi-Fi as well as over the mobile telephone network, and of course it can communicate with you visually over the screen and aurally over microphone and speaker.

In sports, there are devices readily available that can record our every movement. Their set of sensors may contain gyroscopes, compasses, accelerometers, GPS technology, and others. And of course there are a lot of other devices, for example, to monitor carbon monoxide levels underground or in big cities and other dangerous gases and particles in the air.

Currently a team at the University of Michigan is developing a wearable vapor sensor that can continuously monitor patients who have lung disease, diabetes, high blood pressure, or anemia. Also, GM is planning a monitoring device in your car that will make sure you keep your eyes on the road. Soon a lot of devices will be watching every breath you take and every move you make, and you'll love it.

Smartphones, tablets, and computers are the ideal devices to interface those sensors with humans. Maybe in the future your fridge will whine to your mobile phone if it has a problem. If you link a camera and a speaker to your home security system and that to the Internet, then you can supervise and talk to your dog via your mobile phone from wherever you are. One new gadget I came across the other day is a sensor that you stick into the dirt of your potted plant. You can specify the plants needs so the device knows what to monitor, and every time you walk past it, the device will tell your mobile phone what the plant needs.

In 2012, when people at Ericson tried to get their heads around the usability of the Internet of Things, their User Experience Lab teamed up with LEGO Mindstorms to build robots that demonstrated the ideas and usability of their concepts. The idea was that robots or just simple devices communicate with one another to get things done [11].

For example, one robot sorted socks by color and communicated with the washing machine about the right washing program for the different socks. Some soil sensors reported to the watering robot when the plant they were monitoring needed water. The dog-like mail robot Paperboy2.0 fetched the newspaper when the bell rang.

Sure, these examples are toys and not always realistic. In my world, when the doorbell rings, the dog runs to bite the mail carrier. It would have helped if Paperboy2.0 had listened to the mailbox instead. But you get the idea.

In a factory setting, a lot of such functionality could be useful. For example, if your supplier recently delivered a truckload of widgets that you have to build into your products, then someone or something knowing what has to be sorted into your machinery would help a lot. If those widgets could communicate with the system that controls the production process and if they could be localized via VLP, then it would be easy for the system to send the responsible person a message telling her what to fetch next, where it is, and where it has to go.

What is important from a manufacturing and automation point of view is that there are more and more sensors on the market now that have their own intelligence and can communicate with the outside world. Wireless communication of parts and machines could save a lot of wiring costs and provide more flexibility in manufacturing automation.

Chapter 7 | New Manufacturing Technologies

The number of patent applications is growing as the price of sensors is falling. The cost of a basic RFID tag has fallen to about 10 cents (US), and the cost for micro-electromechanical devices like accelerometers or pressure sensors has fallen by about 90 percent. In many automation projects, I would not want the sensors on those machines to communicate via Wi-Fi or Bluetooth with the controller for safety reasons. However, there are protocols now where all communication can run over the one cable, which is used to feed the sensors as well. This means that instead of connecting all the sensors and actuators with power and then with the central controller, in the future there will be only one wired loop that goes from device to device and from sensor to sensor carrying power and signal.

For example, to control the position of a machine head, you could have a micro controller for every axis. There is only one cable connecting all three axis controllers to the main controller. The main controller then just publishes the next position, and the axis controllers read their bit and act accordingly.

And there is more. If the parts in your factory become smart too, then you can do many more things. I'll show a simple example of mass customization here; there will be more complex ones later in this book. Say there is a Make Your Own Soft Drink (MYOSD) manufacturing business in your city. If you buy 12 bottles, you will get free home delivery. And here's how it works.

You go to the MYOSD web site and specify the soft drinks you would like to have. You can choose from a large range of proven recipes, or you can specify your own recipe. It's best if you start from something you know and modify that, hoping to make it better suit your taste. You can give your soft drink a name, and the web app will store your recipe, so when you come back, you can order it again or modify it first, whatever you would like to do.

The hardest bit for this new business will be to make sure you don't order anything that tastes really awful.

Once your order is in (you can order 12 bottles with a different recipe for each), the recipes are downloaded to a chip attached to a bottle. Each bottle will get a chip with a recipe and run with that through production. The bottles will run past all fill stations and be filled with the ingredients specified and then filled with water and carbonated according to the recipe. A label will be printed, showing the name you've given that particular drink and automatically attached to the bottle.

At the end of the line, your 12 bottles will automatically be packed into a box labeled with your name and address, and you receive your soft drinks the next morning at your doorstep.

The exciting thing about this example is that there are no new technologies needed to make it work. All the technologies exist; all it needs is a bit of a software development effort. As far as I know, nobody has tried it yet, but if you know someone who has, please let me know.

Additive Manufacturing

At the Manufacturing the Future business breakfast event organized by the Bankstown Business Advisory Service on April 2, 2014 [6], Professor Goran Roos told this story of a Chinese manufacturer of plastic parts he had the privilege to visit. The managers of that factory told Roos that they had started to produce the tools for manufacturing their plastic parts by themselves, using additive manufacturing. There is a specialized tool-making industry that usually supplies those industries, apparently in China too. I guess those toolmakers work a bit cheaper in China, which explains the figures he gave. An ordinary mold from the toolmaker would cost that Chinese plastics part manufacturer $500. Produce it with a 3D printer costs them now $10,000.

So, why are they doing that? The reason is that this way they can produce a way more sophisticated tool, one that allows them to cool the part much quicker than with the ordinary mold. According to Roos, they claimed that the increased productivity as a result of a reduced cycle time gives them a return on investment (ROI) on the tool that can be measured in hours!

IDTechEx forecasts the global market for 3D printing to grow to $7 billion (US) by 2025. After the current hype around the technology, they expect a steady growth in applications. Over the next few years, the major driver will be the production of parts that can't be produced in any other way, as in the example of the plastic manufacturer.

Prototyping will continue to play an important role as well, and more and more production will be for unique parts or small batches, parts that would be more expensive to produce in a different technology.

The use of additive manufacturing technologies will certainly spread wider as more and more materials become available for additive manufacturing. In that sense, the technology is still in its infancy, but as a manufacturer, you should closely observe this trend to make sure you are among the first to use the technology when it becomes advantageous to do so. In particular, you should be aware of the potential of the technology in terms of what you could produce that is not possible to produce with classical subtractive methods.

Additive manufacturing technologies are already used, particularly in manufacturing parts for jet engines and other parts in the airline industry where the ability to produce better parts is more important than price.

One important advantage of additive manufacturing is that you can produce almost anything, without having to think too much about how it is to be done, once you have worked out the basic principle of how to produce a whole family of parts. This will in the future be a big enabler of individualized production, which I will discuss later in the book in greater detail.

Chapter 7 | New Manufacturing Technologies

At the Composites Australia Conference 2014, Dominic Parsonson presented many uses of additive manufacturing technologies in the composites industry [12]. This goes from printing complex inner cores over which the fiber and resin composite is built. Using a soluble core allows manufacturers to easily remove the core after the composite structure is cured.

Additively manufactured cores can be used as well for mandrels at the end of a standard aluminum shaft. This enables the efficient production of individual ends to single pieces or small batches of composite pipes.

As an example of a 3D printed structure directly used in a composite product, Dominic Parsonson presented some net-shaped cores, that is, self-supporting, high-strength, lightweight polymer cores to support complex assemblies. This is a good example of where individualized production could become possible in the near future.

Other applications are in rapid prototyping and in producing tools for manufacturing of small batches, for example drilling guides or layup guides.

In most applications of additive manufacturing, the biggest challenge is the automation of the specification process. Producing complex parts in the subtractive way was always a challenge in that respect, so it is usually automated only if there are reasonable batch sizes. The specification process was then in that sense part of the process of trying to work out how the part can be produced. With additive manufacturing, however, the step from specification to production is usually fairly easy, especially if the production process of roughly similar parts is already known. Thus, coming up with an efficient specification process in that field is a new challenge.

A good example for this is Invisalign, a teeth-straightening system used by orthodontists. They take an impression of the current state of the teeth and send that to the manufacturer, where it is scanned in. From that their software calculates and 3D prints a set of (almost invisible) aligners that have to be changed every two weeks, until the final state of the desired teeth position is reached.

Robotics

Recent developments in robot technology will have a revolutionary effect in automation. Until now, robots were expensive and inflexible. You could do amazing things with them; however, they had to be programmed or taught and would then just run the same sequence of manipulations over and over again.

For example, in a windshield factory, robots were used at the end of the production line to grind the outline of the windshields. A grinding disc was rotating at a fixed position, and the robot would pick up the next windshield from the production line. The robot would then just run the edge of the windshield past that grinding disc.

Mass Customization

To make this operation possible, everything has to be precise. Worked into the windshield's ceramic print along the edge were positioning points that allowed positioning the windshield precisely for pickup by the robot. The robot then had to be programmed with the precise movements corresponding to the outline of the windshield to make sure there was always an edge of .1" to .2" (.25mm to .5mm) to grind off. Any less and the edge would not be clean; any more, and there would be too much load on the grinding disc, burning the glass.

This meant that the whole operation was pretty inflexible. When changing to the production of a windshield for another car, the whole production line had to be reconfigured. Part of this exercise was that the robot had to be loaded a new program as well.

In his January 2007 article "A Robot in Every Home" in *Scientific American*, Bill Gates wrote that the development of robots was at a similar stage as the development of computers was in the 1970s. Are we now at a stage where computers were in the 1980s? At the end of the 1980s there was a computer in every home. I guess at the end of the 2010s there will at least be a cleaning robot in every home. They have been a long time coming but are pretty cheap now. The same holds for the lawn mowers. It's the increased capacity of batteries that helps drive this evolution, not only here but in so many other fields as well.

If you are too young to remember the computers of the 1970s to see the development since then, Google it. Computing has come an amazingly long way since then. If robots undergo a similar development in the next 30 years as computers did in the last 30, then that means not only an industrial revolution but one of the whole society as well.

Robots have already become way more flexible. They can be programmed dynamically so that a production line can produce a different piece every time. And not only that, they have become much cheaper. That means that general robots can now be used in many more manufacturing processes than just a few years ago.

Examples of flexible lightweight robots are the Baxter Machine or the recently released third-generation UR5 and UR10 of the Danish manufacturer Universal Robots A/S. Among other things, these new robots have a patented adjustable safety system with eight new safety-rated functions, better connection possibilities for the control boxes, and twice the number of built-in I/Os.

The safety systems of these new types of robots allow them to operate without a safety cage, thus allowing safe human-robot collaboration. Universal Robots' UR5 and UR10 are even safety certified by TÜV. These robots are easily programmable without complex technological know-how and can be handled via touch pad.

A good example for how the use of robotics can transform and future proof a company is Rode microphones in Sydney, which went from building 600 microphones a year for the Sydney market to producing 60,000 a month using robotics. It is automated to the degree that production can be kept running over the weekend by one person [13].

Assistive Automation

There are various ways you can improve your productivity. Assistive automation is one of them. This is not about replacing workers by automatons but by assisting them to be more productive. Often people talk about *assistive automation* if they use robots to do part of the job, for example the heavy lifting. I would like to use this term in a broader sense, understanding it as any type of support you can give the worker to become more productive.

From where I come, this means of course using ICT to assist, in most cases in combination with some machinery. Small and medium enterprises (SMEs) in particular need flexible production systems to deliver customized products. In an assistive manufacturing context, the system already knows what the next job is and has all the data to do it.

A recent development consists of providing robots with some perception mechanisms so that they are "aware" of the humans working beside them and act accordingly. The robots become aware of other obstacles too and are able to navigate around them when doing their job.

If done right, the worker will be able to do more complex tasks. In Germany, BMW has used the technology to enable older workers to stay longer in their jobs. However, this can also mean that the worker needs to be upskilled. Not only are robots getting smarter, but the human-robot interface is becoming more complex and needs competent operators.

Upskilling workers might not always be possible, as discussed elsewhere. In some places in Europe, it is becoming more and more difficult to get young people into factories where they might get dirty hands. This is sometimes the driving force behind trying to keep older skilled workers active for longer in the workforce. A study in northern Italy showed that many of the highly skilled and well-paid artisans, working with stone, timber, and leather, struggle to find successors for their businesses and often switch to northern African immigrants.

General-purpose robots become more complex and are able to handle more degrees of freedom, but a lot of effort goes into the programming and the user interface. If you understand a robot as simply being a programmable manipulator, then robots are everywhere, although they do not always look like the ones in the automotive industry.

A lot of work is being done in the field of mechatronics, such as mechanical systems combined with electronics and in particular miniaturized precision systems in that area. Robots can help with the automated production in that field just by uploading the right program code for the product to be produced next.

Thus, you can increase automation and productivity by enhancing the sensing, intelligence, capability, and performance of robotic systems. With enhanced human-machine cooperation you can then achieve higher throughput and precision.

Advanced Composites

As I learned at the 2014 Composites Australia conference [14], people seem to better understand what I am talking about in this chapter if the term *carbon fiber* is used. However, there are many fibers and resins that can be used, and often have to be used, to get the desired properties of a product. Common examples are combinations of carbon and glass fibers.

Recent successes in the use of composites include the commercialization of the first carbon fiber wheel, currently used mainly for expensive fast cars, or a draw-footbridge over a channel in Wales. There are fields where composites are the only way to go, such as in boats of every size and style, from the rowing skiff up to the racing yacht.

Less familiar is the use of composites in cars and planes. However, there is an increasing number of parts used in those fields. Advanced composites are the main material used to build the air frame of the Boeing 787 Dreamliner. Overall, about 50 percent of it consists of built-in composites [15]. Apart from the wheels mentioned earlier, a lot of other parts of cars are composites these days. In many cases, weight is the critical issue (boats, planes); in others, it's durability. Weight is also an issue for car parts but not at any price.

The advanced composites industry is greatly diversified. There are manufacturers that can build complex single pieces and others that are able to produce large batches. By 2020 it is estimated that 75 percent of the industry will be automated [16]. This estimate includes the production of big batches of identical pieces, though.

It is a bigger but not impossible challenge to bring a higher degree of automation into those companies that build single pieces and small batches, in particular those applications where every gram matters but strength also must be guaranteed. An automated approach could make it much easier to produce within those parameters.

There is a wide range of technologies available to mass customize the production of composite parts. There is even a way of additive manufacturing using short fibers. This method, though, does not deliver higher strength to those parts. Automated fiber placement uses continuous fiber–reinforced tape, placed with a robot and heated during placement either by melting thermoplastic tape into it and then rolling it on or by using thermoset tape directly.

In [16], A. Beehag et al. present two research labs dedicated to developing automation technologies for composites manufacturing, one in Bristol, United Kingdom, and one in Germany, but there are more all over the world. While the German center is more aerospace-oriented, the Manufacturing Demonstration Facility at the Oak Ridge National Laboratory [17] deals with more complex composites. An example is an outer metal layer that is 3D printed. This is an approach that would facilitate mass customization.

Nanotechnology

As I said earlier, I'm not an expert of all things manufacturing and can't discuss all new and emerging technologies here, but as a last topic in this chapter, I will talk about nanotechnology because it is widely seen as one of the enabling technologies for the current industrial revolution. Jenifer Khan predicted in 2006 that nanotechnology will evolve over the next few decades in a way that makes the computer revolution look like small change [18].

Nanotechnology is so fascinating because materials behave so differently at the nanoscale level. This is because the nanoscale level is where the essential properties of a material are determined. Depending on how you arrange carbon atoms, you get coal, graphite, or diamonds. On the nanoscale, carbon nanotubes and graphene are most fascinating structures.

Graphene is one layer of carbon built in monocrystalline form. Since it is only one layer of atoms thick, it is completely invisible. If you could build it so big, you could make a hammock out of it, then this one layer of atoms would be strong enough to carry a cat. A lot of graphene research is currently being done in electronics. For example, if graphene is placed on boron and its crystalline structure aligned with that of the boron substrate, then the electrical properties change significantly. In this new material, also called *white graphene*, it is possible to make electrons travel almost perpendicularly to an applied electrical field, thus consuming little energy [19]. That way, it might once be possible to build electronics that use very little energy, bringing us a step closer to the vision of having lots of miniature sensors that collect energy from the environment and communicate infrequently on significant state changes.

Other applications of graphene are miniaturized sensors. Graphene combined with other materials allows the creation of very sensitive sensors for medical and other devices. The incredible mechanical strength of graphene might become useful in advanced composites, and a lot of research is being done currently for various electronic applications. A recent discovery showed that graphene could vastly improve the efficiency of fuel cells and of batteries as well.

Carbon nanotubes could be the new carbon fiber. Researchers try to spin them in long strands that have the potential to become much stronger than Kevlar. They have the potential to become 100 times stronger than steel while only one-eighth of its weight. Simply adding nanotubes to epoxy makes the glue 30 percent stronger. Carbon nanotubes could be teased into twine and woven into sheets. If they were used in composites, it could revolutionize the way we build in the future and the height to which we can build.

The art of carbon nanotubes and graphene lies in growing structures from the bottom up. In the case of graphene, that single layer of carbon can be used to grow electronic nanostructures on its surface. The result would be very thin lightweight flexible electronics. Nanotubes, on the other hand, could have the potential to carry very high loads of electric power with significantly less loss than copper or aluminum.

There are a lot of potential biological and medical applications of nanotechnology too. Direct medical use has its dangers and is probably still more than a decade away. However, research is going on to develop sensors that can detect proteins, for example ones in your breath that might indicate cancer. One other application is the development of nano filters, which have a high potential for the purification of water.

Although most research progress has been made with carbon-based nanotechnology, there are other materials and technologies used as well. For example, silver nanoparticles are used as an antibacterial agent or titanium dioxide in sunscreens. Clothes have been infused with nano particles to make them more resistant to dirt and to last longer. Replacing the anode with nanoparticles of lithium ion phosphate can vastly improve the performance of electrical batteries.

One recent but already commercialized innovation using nanotechnology is the development of supercapacitors. These devices can store very high charges of electrical power. There are already applications to drive trams in China that way. In Guangzhou and Nanjing there are new tram lines with no overhead line but charging stations at the tram stops instead. On a charge of the supercapacitors, the trams can run for about 2.5 miles (4km), and the super capacitors can be recharged at the stops within about 30 seconds.

PART III

Vertical Integration in Mass Customization

> "We can only see a short distance ahead, but we can see plenty there that needs to be done."
>
> —Alan Turing

Vertical integration is about the integrated data flow from design over order processing through to the machines. In an ideal world, the data flow in the backward direction would be supported, too; however, I fear that will take a bit longer.

Let me illustrate with an example what I mean. In the construction world, some twenty years ago the concept of the building information model (BIM) was introduced and has steadily grown since. BIMs are software objects that can be incorporated into technical drawings. In contrast to other technical drawing components, BIMs add more dimensions to the drawing than just their 3D representation. For example, one additional dimension is time, the other cost. So if, for example, an architect adds a BIM supplied by a window manufacturer to his drawing, then that BIM might be able to calculate an estimate for cost and delivery time. This is possible if the pricing structures of the window manufacturer are open and can be reproduced by their customers.

The disadvantage of such a system is that the BIMs have to be updated if prices change. An elegant way to avoid this would be a price calculator in the cloud so that every BIM could ask for its price at any time. Extending that line of thought, this could even mean that there is only one standard BIM for all window manufacturers. That would enable the architect to get current prices and delivery dates from all window manufacturers he works with at any time.

The concept of data flowing backward could then be extended a bit more. After ordering, customers could get feedback on the state of the production process. So, if they have to make changes, they could work out what the consequences are and see as well if there are any delays with delivery.

The BIM concept has come a long way already, and standardization attempts have been made in many areas. Their advantage is that they are passed on from the design team to contractors and subcontractors, who can then add their specific information so that no information is lost and has to be entered again. However, with the large range of BMI objects out there, standardization is complex and takes quite some time.

Vertical integration is crucial in mass customization. Since all products are customized, this means much more data has to be entered than when processing orders for standardized products. Therefore, it is important that all data is entered only once, either by the end customer or by product experts along the supply line and down to the machines. If machining know-how is required at the machine, then of course that part information has to be entered by the operator at the machine, but we have to make sure the operator needs to enter only that specific information.

The next two chapters talk about software integration, first in the conventional way and then about the advantages and problems when moving your applications to the cloud.

CHAPTER 8

Software Integration

"As a rule, software systems do not work well until they have been used, and have failed repeatedly, in real applications."

—Dave Parnass

A recent survey by the Australian institute Commonwealth Scientific and Industrial Research Organisation (CSIRO) showed that the following points are true regarding small and medium-sized enterprises (SMEs):

- Ninety-one percent have Internet access.
- Forty-five percent have a web site.
- Fifty-five percent place orders online.
- Twenty-eight percent receive orders online.
- Thirty-nine percent of SMEs with one to four employees receive orders online.

But even of the 28 percent that receive orders online, it is only a small percentage of those that then go on to process the received order electronically. The usual procedure is that those orders are then manually typed from the screen (or bits of the order are copied and pasted) into the order-processing system.

Research has shown that this lack of software integration can cost up to 10 percent of a company's profit. If you have a lot of small customers, each ordering in a different format such as in an unstructured e-mail, then you could try to set up an ordering form on your web site that would ensure you always get orders in the same format for further processing.

Chapter 8 | Software Integration

If you have a large customer always ordering in the same format, then it should be easy enough to write a small program that imports from there into your system. Even if there are a lot of specialties that need to be attended to, this does not mean that the basic order information has to be entered manually every time.

There are almost always some tricky bits when trying to integrate different software products or modules. However, it is almost always possible to gain a lot of productivity by doing the easy bits first.

Getting the order automatically into your system is of course only the first step of integrating your software systems. Data should then flow all the way through your system. No bit of information should be entered twice. Of course, more information will have to be added during processing. So if your order is in, then you will have to plan for its production. This can mean a lot of new data, most of it generated by the system. It might generate data for all the parts of it, for the production of the parts, for the assembly, and so forth. And of course it becomes included in the production plan, parts and assembly get production dates, and so on.

If you have many different products with many variants and options or even individualized production and if those products are complex and later in their life might need support, then it also helps to keep all of the relevant data in an archive. If in 20 years a problem arises, this makes sure you can help efficiently.

Inefficient data processing can lead to or is a symptom of other inefficiencies, meaning you most likely have high delivery or processing times too.

Processing orders and producing single items or small batches efficiently is especially useful if you have a large variety of products. This prevents trying to hold a lot of different variants in stock and thus binding a lot of capital.

Often this involves investing in software that brings the production data to the factory floor. This could be a simple information system that tells the people what to assemble how or software that controls the assembly line and makes it produce the requested products. If custom parts and components have to be produced, you will get the best efficiency if you make sure all the data you have is loaded onto the relevant production machines too.

Currently most software integration is done by using a database that stores all data for all programs and processes relevant for the products. All programs then have access to all the information they need and that has already been entered. Massive improvements in database technology in the last 20 to 30 years meant that bigger and bigger systems could be supported and that many processes could be automated internally, for example with stored procedures.

Mass Customization

If you have to order customized parts in that context from suppliers, however, then that process is usually not supported that well. If you know the supplier's pricing model, then you can replicate that on your system and thus still automate the quoting process. However, every time the supplier changes prices, you will have to reflect that in your software too, and if you want to change to a different supplier with a different model, then you have to change your calculation software too in order to reflect that.

Once you have your order, you will have to export orders to your suppliers too. Quite often this is done by automatically generating e-mails with the technical data appended. For the supplier, this means that they have to import your order into their system, verify and potentially complete the data, plan the production, and then generate an order confirmation that more likely than not will come back per e-mail.

So if you work this way, you cannot confirm a delivery date to your customer before you have the delivery dates for all externally supplied parts and components confirmed and entered into your system.

This of course is still better than having to enter all the data manually multiple times. But if you manage one of those many SMEs that still do that, then it might be a good idea if you try to figure out first whether you could go into the cloud with that application. Talk to your suppliers and business-to-business (B2B) customers first. If you can't find a solution that's already close to what you need, at least try to find a software developer who understands and has some experience in the technologies involved.

There are some additional problems and issues with cloud applications, particularly with large ones. If you are an SME, then data volumes most likely are not that high, so don't be overly concerned when reading the next chapter. Any form of ordering electronically will have its problems, so that's no different in a cloud solution. Other than that, your database will reside on a server in the cloud somewhere instead of your office, so you don't have to look after it yourself.

CHAPTER 9

Going Cloud

> "Look at a sky in which there are one or more clouds and you will see something no other person has ever seen, since the dawn of time."
>
> —J.D. Boatwood

I do not advocate that you use a free public cloud system for your business data (and I recommend being careful with your private data on those systems). Cloud services are about economies of scale. The central piece of infrastructure in 19th-century manufacturing was power. Every factory had their own mechanical power generator, be it a steam engine or a water mill. Only with the invention of electricity did it become possible to centralize power generation and distribute it to wherever it was needed.

Something similar happens now with cloud computing. Currently there are millions if not billions of computers, with hard disks half full, idling most of the time. Cloud computing has the potential to save a lot of money by centralizing computing power and selling it as a service.

If you are a small to medium-size business and work with a professional cloud service provider that provides a dedicated cloud solution, accessible to you and your employees only, then this not only is cheaper but opens a lot of new possibilities for you as well. Look for a solution specific to your industry. For example, my company offers a solution specifically for small to medium-sized glaziers. There are multiple advantages to such a system.

First, you can access the software and do everything you need without having to install anything. It just runs in the browser. Second, you can connect directly to your business partner, for example your suppliers, from within the system. If you are a small installer, then you can calculate your quotes there, and the system automatically factors in your suppliers' prices and delivery dates. If you change a quote into an order, then your external pieces are ordered automatically. Furthermore, because you have access wherever you have an Internet connection, you can update the system from wherever you are. Updating the status of a job to "installed" could easily be done from a smartphone, before even leaving the site. If you want to enter quotes in the field, I recommend using something a little bit bigger, such as a tablet computer.

If you are a big business, you can run your own server farm and have a cloud solution on your intranet, where you can process almost everything. The big advantage of this is providing all your employees, including those on the factory floor, with access to exactly the information they need. Then, all updates go directly to the central database.

Big business was until recently using cloud services more at the fringes of their business. However, things are changing fast. I can see more and more functionality going into the cloud. Companies start moving their whole systems there, and the solution providers are right behind them.

Most start with customer relationship management (CRM) since all the support you can have on the go helps your sales. What then follows is order processing because there is already data from the CRM that can be reused. The logical next step is production planning, followed by the whole enterprise resource planning (ERP) system, and soon all of your business is in the cloud.

Many don't want to go down that path yet and move to the cloud only what makes most sense to them. Whichever path you choose, there is a lot of work to be done. With every part you move to the cloud, there is an interface with the rest of your operations that needs to be created.

In this context, I consider a cloud solution as having one or more large server farms somewhere in the cloud that store all of your relevant data. Whether there is any other communication or load sharing between devices on that network is a secondary consideration.

In addition to order processing you want to run all you can for your production in the cloud as well, which is all but the time-critical parts. In doing so, you look to capture all the advantages mentioned earlier.

Therefore, I will discuss some issues that arise with large cloud solutions. One of the biggest issues is of course security, but I do not want to get into that here; the fact that I'm not an expert in that field is only one reason why. The other is that these issues are strongly dependent on the solution you choose. If you work with a solutions provider, then you will have to discuss with them the necessary security measures. They will implement what you need, and you will have to stick to the protocols they provide. If you run your own cloud solution, then it is your IT people's job to keep your data secured; make sure you have the right people on that job.

Another issue that arises with large cloud solutions, though, is the problem of data consistency. To illustrate this, let's start with a classical story from computer science.

Byzantine Generals Problems

The classical Byzantine generals problem deals with the problem of figuring out if one of several Byzantine generals is a traitor. In computing terms (and in its simplest version), you want to determine on how many computers you need to run the same calculations to figure out if one failed and which one it is. In that simple case, you would need four generals (computers) if you want to reliably identify a traitor (defect computer). In the general case, you need more than two-thirds of the generals to be loyal to be able to detect the traitors.

Here I will concentrate more on another problem, the two Byzantine generals problem, which is less general but older than the other one. Two Byzantine generals are at war with a common enemy. The enemy is sitting in a well-defended fortress at the top of a hill. The two generals are camping with their armies at the base of the hill, at opposite sites.

The two generals can win only if they attack at night at the same time. In a meeting they agree to wait for the right conditions and then to attack at 3 a.m. Both generals can initiate the attack if the conditions seem right to them. The other general, though, has to confirm that the conditions are right for him too.

But there is a communication problem. The enemy has patrols out, and they try to disrupt communications by killing or capturing couriers. Thus, if one general plans to initiate an attack but does not get confirmation, then he does not know whether his message was received and will not attack since on his own he does not have a fighting chance. If he does get confirmation, then he could attack.

On the other hand, if the confirming general does not know if his confirmation has been received, he cannot attack because he risks attacking on his own with the same disastrous consequences. Because there is no line of sight between the two camps, there is no other way of confirmation.

So, what could be done? If the initiating general sends a confirmation message, then he does not know whether his message gets through, meaning that it's not safe to attack and so forth. Is there a solution to this problem? We can't go back in history and check what the two Byzantine generals did because no such problem is known from history. The problem was simply devised to illustrate a puzzle. In the 1990s, it was proven that there is no solution (see Figure 9-1).

Chapter 9 | Going Cloud

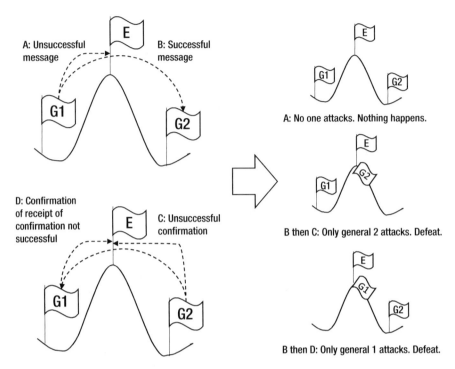

Figure 9-1. Two Byzantine generals problem. Because all communications have a chance of failing, a general who receives a message can't know if his confirmation was in turn received, potentially leading to disaster.

The Two-Phase Commit Protocol

So, why am I telling this story? People started studying these problems when the first wide area networks came into play. Let's look deeper.

Assume for now that a friend lent you this book to read because he thought you might be interested in it. It turns out you are so convinced, you want to buy the book for yourself. Thus, you go to Amazon, order the book, and pay by credit card.

Let's now analyze what your credit card payment implies. A central server, initiated by Amazon, will try to process that payment. To do so, it will start a transaction by requesting your bank confirm that there is enough credit available to buy the book. Furthermore, it will have to make sure the receiving account is available and ready to receive the money. Only then can the transaction take place.

In reality there is much more going on since your bank will be debited by the credit card organization and Amazon's bank credited. The credit card organization will deduct a processing fee as well, and Amazon has to mark one of my books as sold and send it to you.

To make sure this all happens properly or not at all, what is in widespread use now is a protocol called *two-phase commit*. The initializing computer announces the transaction to all other computers and expects a yes or no answer from all of them. The receiving computers will test whether the transaction is possible and, if not, respond no. Otherwise, they set up their transaction and log them. The affected records on the database remain locked.

Once the initializing computer has received a "yes" by all computers involved, it will send a "commit" message to all of them. These will then access their logs and process the commit operations, which in general consist only of releasing the locked records.

If the initializing computer receives a "no" by one of the computers involved, it sends a "rollback" command to all other computers, and they perform an undo operation, taking back all they had done before, based on their logs.

If you understood the Byzantines generals problem, then it must be clear to you that this will not always work. If, for example, all except your computer receive the commit, then the transaction goes through, but your credit card record, together with a few others of your bank, remains locked. At some point in time, for example the next time you want to use your credit card, it might be blocked. It is not clear how much credit is left. You will have to call your bank, and someone there has to analyze the log and then try to figure out what should happen, the commit or the rollback. This can be resolved over the phone with the initializing bank, which is over a second channel of communication, something the two Byzantine generals did not have.

The CAP Theorem

The reason I'm telling you all this is to explain that computer networks have their problems, and sometimes there is no one to blame when something goes wrong. To dig a bit further, I will explain another problem in this space that was only recently investigated because the problem is new. This problem arises if your application is big and accessed from all over the world. Examples are Google, Facebook, or, to get back to our previous example, Amazon.

To provide better access and shorter reaction time, Amazon runs several computer clusters all over the world. Those computers all know how many examples of a product are on stock on the only site that has stock in the United States. If you order a book with Amazon, then your local server sees that the book is in stock and lets the transaction proceed. During this transaction, all other servers will have to decrement the number of books in stock. I don't know how Amazon does this, but one way would be that the decrement is broadcast when the transaction goes through at the main site.

Because this all has to operate over Internet, even if Amazon has a dedicated network, this can take a few seconds. Now let's assume that this book is really good and you are not the only one who wants to buy it. Many have gone before you, and there is only one book left at Amazon. You may sit, say, somewhere in Europe, and another guy in Australia wants that book too, at the same time you try to buy it.

What can happen now is that your server in Europe and the one in Australia could go through in the same short time interval with the sale of the one book that's left. Let's assume that your sale goes through first at the U.S. server. This means that when the sale from the Australian server comes through, then there is no book left anymore, and the poor guy in Australia misses out. This is quite puzzling to him because everything looked good; there was one book left, and he managed to order it before anyone else seemed to get hold of it.

When these types of problems started to emerge, Eric Brewer, a scientist at the University of California, Berkley, postulated a conjecture at the 2000 Symposium on Principles of Distributed Computing, stating that a distributed system like the one described earlier cannot guarantee the following three properties at the same time [20]:

- Consistency (that is, all nodes see the same data at the same time)
- Availability (that is, every request receives a response about its success or failure)
- Partition tolerance(that is, even if connections or messages are lost, the system continues to operate)

In 2002 Seth Gilbert and Nancy Lynch proved a slightly modified version of the CAP theorem [21], and for the next ten years most in the industry took the somewhat simplified view that you can have only two out of the three at the same time.

If looked at that way, the choice is between consistency and availability because in most cases you don't have the option to drop partition tolerance. Because availability was the reason for the partitioning in the first place, consistency has to go. The best achievable under these circumstances then seems to be what is called *eventual consistency*; that is, if no node is isolated, then eventually consistency will be achieved, provided there is no further input to the system.

What *eventual* means depends largely on the size of the system and the information that has to be updated everywhere once input has stopped. Consider Facebook. When you update your profile on Facebook, you are always connected directly to the master server. Once you have updated your record, you stay connected to that server for another 20 seconds before you are routed back to your local server. This is based on the assumption that 20 seconds is plenty of time for your local server to update so that when you come back, you still see the information you've just entered.

Now let's look at another example. You have two servers serving two different areas with the same data. Every time one server updates a data item, it has to send an update to the other server. Now let's assume server A sends an update to server B but can't reach it either because server B is down or because it is separated from A's network. If, when server B comes up again, it can't reach server A, then what you see on server B is stale data. Making things worse, if server B now updates the same data item with different information, once the servers manage to reconnect, the servers have contradicting information and update requests.

In the February 2012 issue of *Computer*, Eric Brower stated that the CAP theorem is widely misinterpreted [22]. First he makes the point that partitions are rare, and if there is no partition, then there is no reason to forfeit availability or consistency. In practice, this means you should try as hard as you can to avoid seriously partitioning your system. For example, although a server farm forms a network that can serve thousands of users, a partition of the individual servers would be a serious defect and hardly ever happens. Only if you have multiple servers in different places do occasional partitions become likely.

The second point Brower makes is that there is not simply strong consistency or eventual consistency with nothing in between. There are engineering trade-offs, and for every instance the right solution has to be found. One of Brower's examples is the network of ATMs of a bank. You would expect that in that case strong consistency is the right choice because otherwise an ATM can't know if there is still enough money in the user's account. But in practice, availability trumps consistency. As Brower explained, the reason for that is simple. Higher availability means more profit. Instead of strong consistency, the maximum withdrawal amount is limited, thus limiting the risk for the bank.

In that situation, when the partition ends and the ATMs are reconnected, then a recovery is easy because the operations are exchangeable, and all that has to be done is to sum the withdrawals. If the customer becomes overdrawn, then a fee and interest are charged, and in the vast majority of cases the money will be paid back.

Resolving conflicts after a partition has ended is not always as simple as the ATM case, from a programming point of view. For example, if two servers of Amazon sold the same last copy of a book during a partition, then a recovery would inadvertently mean that one customer would miss out and the money has to be paid back.

More recent research has found that in most contexts networks are not that unreliable at all. Usually partitions are short so that the choice of eventual consistency is quite a good one. This can be the case even when you run a second server in a remote location just to mirror the first one as a backup. Although it seems obvious at first to choose strong consistency, I advocate eventual consistency because otherwise the latency times on the primary server would likely almost double. In the unlikely event of a mayor catastrophe at the primary server's site, a small loss of data would be the least of your problems.

The lesson learned from all that theory should be that as far as consistency is concerned, there are always trade-offs to be made, and they all have their price. Running multiple servers for the same app after all is done because you want to increase availability. If you then request strong consistency, then that is counterproductive. So, there will be remaining consistency issues in all systems you might encounter, be it as a user on the Web or with your own cloud solution if it is big enough to require multiple servers. Please be aware of this, and while you shouldn't tolerate frequent problems, don't be surprised if something goes wrong occasionally.

The Extended Cloud

Mass customization requires suppliers that can mass customize too, and it requires customers that can specify electronically and in detail what they need. In the lock-and-key example, customers are locksmiths. They know what lock and what door the cylinder has to fit, so they can specify cylinder types. They sell to the end customer and specify together with them what keys should open which doors.

To manage that, they use master key management software. From those software packets they are able to export orders in electronic form and send them to the manufacturers by e-mail or upload. Because they have their own prices for keys and cylinders as well as the other hardware they sell, they are able to generate quotes to their customer from their management software as well.

This is the older approach and is how it works in many industries. In the glass example, window manufacturers usually run standard software packages to manage their orders and production, and these packages can export the specifications for the insulation glass units (IGUs) needed in electronic form too.

However, what I believe is that we have to span the net wider and to have a more modern approach. And that is where the cloud comes into it.

Double Data Entry

Just the other day a friend told me about his visit to a huge joinery operation in China. They showed him one of their customer service centers (they have more than 1,000, all over China). Customers come to those centers to pick and choose, for example, the kitchen for their new apartment. They go through the huge showroom and select the style and color as well as the devices and anything else they might want to have in their kitchen.

Next, they come to the desk and flesh out the nitty-gritty details of that new kitchen. The assistant simply asks them for their address and apartment number, and the floor plan shows up on the computer screen.

In the western world, I don't think customers would be too happy if anyone could look up the floor plan of their new house or apartment on the Internet. But perhaps the architect or builder could export segments of the floor plan (in this case the kitchen) and send them to the owner. Then the owners are in control over who has access to the data. Since there are good standards for storing technical drawings, the kitchen manufacturer's software should import these plans easily.

The same applies to your new windows. Someone at least has to check what the new building standards are. Depending on the climate zone, sometimes the orientation of the window matters. There are a lot of things that have to be attended to that most people are not aware of. After breaking an IGU of a window from our fairly new extension, I was able to sort out with the window manufacturer what the specification of that unit was and could thus organize what to order over the phone with the local glazier, without them having to come to me. But if that information was available in the cloud and accessible for the glass community, I could just have explained to the glazier exactly which piece of glass from which window in my house was broken. His system would be able to get the exact specification of that unit including the manufacturer. He could automatically get a quote from them and thus generate a quote for me without entering any data at all.

Vivek Kundra related another example in [23]. It was about an issue noticed with student aid applications. Students used to drop out of this really complicated form. Kundra noticed that the form was asking a lot of the same questions that the students had been asked before, and although simple data sharing was not possible because of privacy legislation, using smart technology Kundra still managed to eliminate 70 questions without violating the law.

In a manufacturing context, data should never be entered twice, but sadly it is frequently entered even more often than that. It is not uncommon that the drawing of a piece is printed out from some software and then sent through to a supplier, who enters all the relevant details into his system. It might not stop there. For example, if the order processing software is not capable of handling all of the manufacturing relevant data, then a copy of the drawing goes on paper to the factory floor, and all the relevant data is entered yet again on some production machine.

There should be one point of data entry, and that is when the order is created, preferably by the end customer. If technical know-how is needed that the customer does not have, then try to automate that part of the system. If there are rules, programs can be written to enforce them. Just make sure the customer can and must specify all data relevant to her. If all of the necessary data is generated from there, then that makes sure the customer gets exactly what she ordered. This remains true even if complex decisions have to be made by people on the manufacturer's side.

Extending the Cloud

I've talked about having all your data in the cloud and then having access from wherever you are. For a sales guy, this might mean wherever in the world he is, provided he has Internet access. For a more processing-oriented internal business setting, this might mean only wherever you are in your business's premises, provided you have access to the Internet.

In the previous section, I expanded to the wider world what has happened within strongly regulated supply chain management systems for a long time in some parts of the manufacturing industry. This is best known in car manufacturing, but in other industries many have tried to achieve similar outcomes as well. As to some extent already described, on his plan the architect specifies the size and type of the window and orders that electronically from the window manufacturer. The manufacturer in turn adds information to be able to produce or order the parts needed for that window; the IGU manufacturer then adds what he needs and orders with his suppliers what he can't produce himself.

But let's now draw the net wider still. Your server in the cloud is not the only one there. Your suppliers and your customers might already be there as well. Talk to them! Maybe they run software that is already capable of exchanging data with yours. Or one of you can easily adapt. The wider the net and the more suppliers you can automatically connect to, the better you can perform. If your suppliers' systems can automatically generate quotes, including a delivery date, then your quoting system can query several competing businesses automatically and come up with the right supplier for that occasion.

Case Studies

In this section, let your imagination run wild and imagine two not so unfamiliar scenarios and how you could expand on the possibilities.

Imagine Glass

Imagine you are a small glass processor in a big network of suppliers, partners, and customers. Customers can be any of the following:

- End users ordering replacement glass from your web site
- Cabinet makers ordering glass backsplashes for a kitchen or mirrors for a cabinet or a wardrobe
- Window manufacturers ordering IGUs
- Builders ordering showers
- Other glaziers ordering from you what they can't produce themselves

Partners are other glaziers in your network. They are sometimes customers and sometimes suppliers if you order from them what you can't produce yourself.

Suppose you are the guy building IGUs in your network. If one of your partners needs an IGU, he might order that from you. Because the unit is for a door and he knows what he's doing, he will order toughened glass.

You don't have a toughening oven, so you order the toughened glass from the respective partner in your network (and that might well be the guy who ordered the IGU from you!).

Until now, everything happened electronically. Because the system can automatically request quotes, it can automatically ask for prices from suppliers, and thus your partner that needs the IGU will immediately and automatically get a price from your system, with which he can quote. If his customer accepts, then the rest will roll out automatically. The system will accept the quote for the toughened pieces and order them. Because it's the year 2025, there is an automated car used for deliveries in your partner network. When the toughened pieces are ready, a transport order is generated in the system, and the transport is automatically and optimally scheduled.

When the delivery truck arrives at your ramp, you unload your stuff including the toughened pieces, and they go onto your sealing line. When the delivery truck arrives the next day, your IGU goes on it, and the truck goes autonomously to your partner who ordered it.

Imagine Kitchen

Imagine you want a new kitchen and there is a web site where you can enter everything, including the following:

- The measurements of your kitchen
- The doors and windows
- The color of the floor and the walls
- The placement of your cabinets and appliances
- The sizes and types of your cabinets, the number of draws or shelves, and so on
- The material and color of your benchtop and the kitchen front
- The material and color of your backsplash

Once you enter the main data, you can hit the display button, and the system asks you to put on your 3D glasses. The system then downloads the model of your kitchen and displays it in 3D. Just by movements of your hands, you move the model so you can see your kitchen from different viewpoints and perspectives.

Next you can go back, change the details, including the colors, the arrangements of the cabinets, and anything else you don't like. You could flesh out more details as well, such as the brand and type of your appliances. And at every stage you can hit the display button again, and the system shows you how your new kitchen will look. You will have your own 3D printer anyway, so it will be simple to print a model of your new kitchen once you are satisfied with its design.

Once you have given enough detail, you can ask for a price. You might even be able to ask for prices from several manufacturers near you. They would all give you an earliest delivery date as well. The system would give you a price for everything you have specified in enough detail, but everything you want to order separately you can just tick off and then hit the order button. Next you give your details and pay a deposit, and your order is placed.

You might not become aware of this, but the most amazing part is what will happen next. The server that holds all the data for your kitchen will start processing your order. No human intervention will be needed. The server generates all the necessary data. It calculates the positions for the hinges and generates the routing code for the cutouts. It calculates the size of all the parts. It calculates the placement of the draw runners and sets the positions for the screw holes.

Next, all the parts are sorted by material, and a layout is generated and optimized for every type of material. All the data is sent to a router, and an automatic feeder will send stock sheets to it. The router will cut the stock sheets with all the necessary cut outs and holes. Labels will be printed and automatically attached to all the parts. Once cut, all the pieces cut on that stock sheet will be pushed off the router and then sorted into boxes by a robot.

All the hardware needed will be commissioned from an automatic stock system and automatically delivered into the box too. At the end, the box will be closed, and an automatic, driverless forklift will load it onto an automatic driverless truck, which delivers the box to your doorstep on the delivery date.

If you ordered that kitchen with installation, then a joiner will show up and start building your kitchen.

Now just for a moment imagine that all the tradespeople in your area are well connected and have commissioned and built that web site where you ordered your kitchen together. That means you could specify not only the kitchen but everything that goes with it as well—all the appliances, the sink, the taps, everything. That means the holes and cutouts in your benchtops are correct, the place left for your dishwasher has the right size, the one for the oven or the microwave is correct too, and everything has the correct mounting holes.

Because you ordered everything there, the system also knows where the overheads and range hood are, and it works out the size of the backsplash. You will have to specify where the cutouts for the power points and light switches will be, and then the order goes automatically to the glazier.

Again, all the data for cutting, polishing, and cutout routing is generated and directly sent to the machine, and with little or no human intervention, the backsplash is produced, delivered, and installed. The same goes for all the appliances, which will be delivered to the installer, be it the plumber for the taps and the sink or the electrician for the wall oven and the induction cooktop.

Remember the first example? If you want to order a new bathroom as well, with a mirror cabinet with a lock on it to store your medicine, all you have to provide is the code for your master key system and the identifier for the cylinder. The joinery app will automatically order the mirror with a hole in it at the right position for the lock cylinder from the glazier and the right cylinder type from the lock manufacturer, and your cabinet will come with the mirror and lock fitted.

Does that sound farfetched? The sophisticated 3D user interface might still be a few years away and the driverless cars for delivery a few years more. Other than that, all the technologies are there. In fact, we have developed most of those applications before in some form or other. All that's left to do is the integration of all those currently separate systems to make it all work seamlessly. As far as the information and communications technology (ICT) side of it is concerned, we are working on that wherever we can. The software development side of the problem is actually the much easier part of that exercise. Bringing partners together is often much harder. Thus, all that's needed is a bit of initiative, and you can make it work!

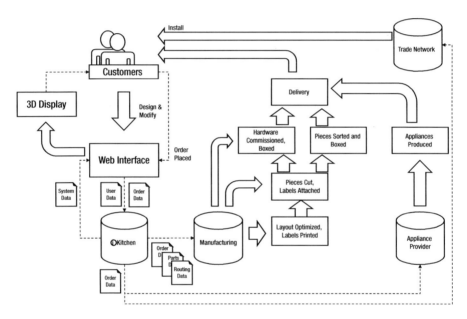

Figure 9-2. An almost completely automated kitchen design solution, from the point of customer specification and order to production, delivery, and installation of the finished product

PART IV

Making Mass Customization Happen
Project Design and Management

> *"If you want to build a ship, don't drum up the men to gather wood, divide the work, and give orders. Instead, teach them to yearn for the vast and endless sea."*
>
> —Antoine de Saint Exupéry

Throughout this book I've promoted agility, and in this last part I provide a structure for an agile implementation of mass customization. Unfortunately, I can't go into too much detail about that process because it will very much depend on the circumstances of what you want to achieve. If there is a big and complex machine at the center of your project, one that will cost more than half of what you plan to invest, then agility will be hard to achieve. But if you have the chance to start small and implement in multiple iterations, then this is the best way of going about the whole project.

Before I go through the process of implementing a mass customization project, however, I will address two important aspects where I frequently have seen things go wrong.

The first is about system performance. If you plan just a simple thing as a first step, understand the complexities of that step well, and know how to handle them, you might want to skip Chapter 10. However, if you plan to rebuild your whole operation in the end, then it might be a good idea to try to understand how it all is supposed to work together by then.

If you plan to start on the vertical path and want to build a tool for your customers so they can specify and order online, then you need to understand the complexities of that only. So, for a start, it will suffice to read Chapter 10's "Sophisticated User Interfaces" section to start. However, for the long run and for your own benefit, I highly recommend that you read the whole of Chapter 10.

The other matter I will talk about is commissioning. Commissioning a new software system on its own is hard, and I have learned many lessons even before printouts, screens, and barcode scanners appeared on the factory floor. If you implement your project in an agile manner in multiple iterations, then you will be faced with commissioning issues at the end of every cycle. Therefore, and because there is a lot to say about that, I will talk about it in a separate chapter, Chapter 11.

CHAPTER 10

Optimizing System Performance

"An algorithm must be seen to be believed."

—Donald Erwin Knuth

Solving Complex Engineering Problems

One of the principles you learn in Lean Six Sigma is to reduce complexity to reduce mistakes. However, while this principle still holds, products and production methods have become more and more complex. So even in mass production complexity will increase.

One way Lean Six Sigma recommends to reduce complexity is to reduce product variability. But if you want to go into mass customization, you need to embrace product variability.

So while I still strongly recommend keeping everything as simple as possible, I also recommend that you don't try to make things too simple. To some extent you have to embrace complexity. So let's have a closer look.

Complexity

We are in the early phase of a technology revolution. It was predicted [10] that by 2020 there will already be 50 billion networked devices, machines, computers, phones, sensors, actuators, routers, and so on, all communicating with each other. Just from an information and communications technology (ICT) point of view, this equates to a software revolution.

In a car today, 50 to 100 controllers talk to each other. In the near future, cars will start talking to each other. They will become autonomous. The number of connections explodes, and we will need to cope with that. But how?

Earlier in the book, I stressed the KISS principle: keep it simple, stupid. Faced with that exponential growth of interconnected devices and an even stronger growth of connections between them, how can we KISS?

In fact, the car is a good example on how to manage complexity. Even in today's cars, not all controllers are equal. Some have higher priority, such as those concerned with safety issues, than others. Separation of concerns is an important way to stay in control and to be able to manage complexity. If you extend that and think of cars becoming self-driving and talking to each other, there will be a governing system that tells the car in every instance where to go or when to stop. But all that cannot override the safety-relevant features of the car and does not interfere with any other functions managed by controllers, for example the radio.

This has not been achieved yet throughout your new car. Chances are that if there is a problem with your radio, then your navigation system will stop working as well. This need not be the case. Just because the two devices share some components, for example the screen, does not mean that if one component unique to one system fails, then the other system has to fail too.

So, not only does separation of concerns help reduce complexity, but it will increase reliability as well. But all that will never make cars as simple again as they were back in the 1950s when the only electronic device in it was the radio.

So, while the KISS principle still holds, you can only try to make it as simple as possible. Anything simpler than that would no longer work properly. And that is where embracing complexity comes into it. Even if you keep things as simple as you can, the complexity of those systems will rise dramatically.

In ICT, this type of revolution used to happen every 10 to 15 years. My first steps in programming were in Fortran 2 on an IBM mainframe. I went through all the generations of systems and programming techniques, from minis, workstations, and PCs to mobile devices. Now it's the cloud, distributed systems, and networked devices.

Mass Customization

For the young programmers in my team, this type of revolution will be a first. They can't remember a time when there were no mobile devices and no laptops. But what's new to this revolution is that it affects everybody, not just people in ICT. It affects all of manufacturing. If your products need not become smart, then at least your production will.

And it's not only the cars. The entire infrastructure around us will change. The way new products are developed will change. The old waterfall method no longer works. Not only software development, but a lot of other industries work different already. After designing on the computer, you have a digital model. Thus, you can simulate and evaluate on the computer too, identifying problems and improving your design in an iterative process until you think you are right. Then you build a prototype directly from your data. Again, you can evaluate and go back to the drawing board (which is now a computer) and improve, simulate, and evaluate. Throughout the process you evaluate not only functionality but manufacturing costs as well. Some industries have worked this way for decades; others have only just started.

You have to become lean and agile. Your competition will, and if you are not as quick as them, you are left behind. This will affect your product development and your investment decisions.

There are risks involved. Higher complexity means a higher potential for error conditions that you can't anticipate. In 2014 alone millions of cars were recalled because of software bugs. You must make sure the high pressure on time to market does not lead to cutting corners. If there are no safety or security concerns, a way to mitigate risks is to make sure you can update software remotely later, so if errors come up you can easily fix them. If they could do that for a rover on Mars, why should that not work here on Earth?

I will discuss these issues in this chapter: evaluation and simulation of complex plant configurations, local and plantwide optimizations, and some thoughts on managing complexity.

The Problems

When planning new investments, you will be interested in the overall performance of your new or modernized plant. In the past, people were optimizing every major processing step, hoping that would give them optimal overall performance.

When trying to automatically produce custom products down to lot size one, a few new unknowns come into the equations. Of course, such a production can run smoothly and profitably only if there are no or negligible setup times. But this is not always the case. Often, what you can get out of the system are reduced setup times if you produce similar products in sequence.

In the kitchen manufacturers' example, the material that usually has to be fed separately is the material for the kitchen front. Bigger manufacturers have a stock area where the most common materials are stored and can automatically be fed to the router. However, less frequent materials have to be loaded separately. So, if you can process jobs with the same material in sequence, you can save setup time for those jobs because you have to load that special material only once. At the same time, if you can optimize multiple jobs together, you have the chance to get higher yield and thus can save on what is usually the most expensive material of the job. Together with researchers from Deakin University, we have done some research on that; see, for example, [24].

In glass applications, the applicator of the spacer frame on the sealing line can hold only a limited number of materials. It takes an experienced operator about ten minutes to change this material. It is therefore a good idea to run all of the day's production of the insulation glass units (IGUs) with the same spacer width and color in sequence. Usually the operator can prepare the next set of materials while the current ones are running, so there is an interruption-free transition to the next type of IGUs. But this is not always good enough. For some infrequent materials, there are short runs. This means that during that run there is not enough time to set up the material for the next one. This led to the idea of placing these short runs in the middle of a long one. When the long run starts, the material for the short run is prepared, and the applicator can switch over to it. At the end of the short run, we are back to the second half of the long run, its material is already there, and we have a smooth transition again.

If your operations become complex, then it will no longer be easy to understand if and how your plant will perform under the various scenarios you can think of. You might not even be sure about what all the scenarios might be.

Because you know your product range, it might be tempting to build a plant that can do it all with the highest possible degree of automation. But before you do that, you have to make sure your system stays productive and you can keep a high yield on expensive materials.

It seems obvious that you should not invest if you don't thoroughly understand how your new plant will perform with that investment. However, I've seen several instances where that due diligence was not done, usually with the consequence of reduced functionality or additional expenses to fix some of the issues. The reason why lies in what I discussed earlier. If you have the wrong people in charge of the project, then you will not get a good outcome.

In many cases, the people in charge specify the different parts of the project without properly understanding how they will all work together. If you buy different machines from different manufacturers, you can of course specify the required performance for every machine. The machine manufacturers will cooperate to the extent they have to, usually to make sure pieces are properly handed over from one machine to the next.

Mass Customization

If the system integrator's job now is simply to prepare the jobs and the data and to control the machines so the jobs can automatically flow through your new factory, then there is one important piece missing: nobody ever checked whether the whole thing is working efficiently under all possible load scenarios.

If the system is easy enough to understand and if the people in charge know well what they want and in what cases they can compromise, then that can work out very well. However, in my career in this business, I've seen that only twice.

On the other hand are the very complex cases, where it is difficult to understand the different load scenarios and how the system will perform. Often it is not clear how to even run the system under special load cases. Again, a simulation can help here.

On rare occasions there is a comparably easy way out of all the complexity. The only one I know is again our lock and key example. When in the early 1990s workstations became widespread and affordable, I built a workstation solution for that. Essentially this was only a much improved user interface to do the same old job in a more efficient way. It reduced the calculation of a big new master key system from about one day to about two hours. However, it did not resolve the problem of later changes of the system, when the customer wanted to extend. In those mechanical systems there was just not enough flexibility, and as a consequence the customer often had to change multiple existing cylinders and keys in addition to the new ones he wanted.

The breakthrough came with a technical invention where it became possible not to have significantly more bolts in a cylinder but to have significantly more places for them and to encode them all on the key. It's a bit complicated, and I can't go into all the details here, but the effect was that customers could set up a hybrid system with some hierarchical structure while still maintaining some flexibility.

The whole structure can be described in the labeling of the keys and cylinders, and once that is specified for a new system, it takes an experienced coder only about ten minutes to specify in a pick-and-choose interface how this is encoded. The flexible parts can then be specified by the customer by just picking the cylinders the new key should open (on the selected hierarchy level) and giving that pattern a name as part of the key's label. From there the computer can automatically calculate all the codes needed.

This has vastly simplified a cumbersome extension process into a fully automatically one, where the customer can specify the new keys and cylinders, including in part their operation mode in their master key management system, and then order electronically. Once the order is imported into the manufacturer's system, all the codes are calculated automatically.

Chapter 10 | Optimizing System Performance

The Need for Mathematics

As a mathematician, every now and then I'm asked what mathematics is used for these days since we have computers to do it for us. I remember quite well, back in the 1970s, when I was asked this the first time by a doctor stitching up my left eyebrow after I had a little accident with my sailing boat. He clearly felt that electronic pocket calculators had made the study of mathematics unnecessary.

In an attempt to explain to him that mathematics happens on a slightly higher level of abstraction than what pocket calculators can do, I told him that I was right now trying to get my head around topological spaces and that as far as I could see there were no numbers used throughout the lecture notes I was working with, except for the chapter numbers.

Math Matters

Every now and then someone does a study on the impact of mathematics on computing efficiency. The studies are usually based on a series of practical problems that take a long time to compute. Moore's law states that computers become roughly twice as fast and powerful every 18 months. After 15 years, computing power thus doubles ten times, and since $2^{10} = 1,024$, one would expect computers to be able to solve those problems 1,024 times faster. But it was long known that in fact these problems could be solved more than 1 million times faster. An in-depth study by Robert Bixby [25] conducted from 1988 to 2004 showed a hardware speedup factor of 1,600 and a software speed up of 3,300, resulting in a total speedup of 5,300,000. With the hardware speedup only, a computation that would have taken two months in 1988 would still take about one hour in 2004, while in combination with the software speedup this was actually done in a one-second snap. The magic behind the software speedup is called mathematics!

About a year ago I came across two new studies of that type, both done for a 15-year interval based on a fairly new problem, sequencing the human genome (or any genome for that matter). The first showed an improvement by a factor of 7 million times, the second by 49 million times. This is reflected by the fact that in the 13 years since the completion of the Human Genome Project the price for analyzing the complete human genome has dropped from $3 billion to about $1,000 according to Francis de Souza, president of Illumina [26].

When Alan Turing (the Bletchley Park code cracker guy) in 1938 invented what we now call the Turing machine [27], he was thinking of a human being when he talked about a computer. Back then, some very time-consuming calculation processes were already known, for example to calculate lens systems for cameras and other optical instruments. These calculations were done on forms that were circulated in a room full of "computers," professionals who always did the same calculation step on a form (but with different figures), and then passed it on. Such calculations could go on for several months.

I remember when I was in charge of my first big software project and got my first minicomputer in the 1980s (a VAX); I went to a course to learn how to handle it. There I met a couple of engineers saying that on the new machine they could do their calculations of optical systems in 4 hours instead of 30 days. Given that this calculation could tell them only that the newly designed optics did or did not do exactly what they wanted, this was great progress. Now they had three chances a day to tinker a bit and calculate again, a process that would have cost them three months before!

What you can learn from this is that as computers evolve, algorithms have to evolve and grow with them. And in most cases mathematicians manage to do an amazing job!

But still, you might ask what does mathematics have to do with manufacturing? All right, there are some high-tech products out there, but once they are constructed, maybe with the help of mathematicians and certainly with the help of a lot of software originally developed by mathematicians, what do you need a mathematician for now?

And yes, you might be aware of some computing-intensive problems that run every day in your factory, such as production planning or layout optimization. With all the new computing power, these old applications now run 1,000 times faster, so what's the problem?

There is a class of calculation problems for which no efficient algorithms are known. These are mostly practical problems, not just academic constructs. A lot of them arise in operations research. Most production planning and layout optimization problems are examples. Because exact solutions for those problems can't be found, heuristics are used in most cases with only a few exceptions for small problems.

What does that mean for you? Remember, over the last ten years not only has computer speed increased, but algorithms have improved as well. It might well be that there are programs out there that not only run much faster but can give you a better production schedule or a layout with less waste. In fact, I find more and more common practical medium-size problem instances, where exact algorithms or hybrids can now be used, thus giving a much higher degree of certainty that the solution is at or close to the optimum.

And don't forget the simulation model you want to do. Mathematicians are good at that sort of thing. So, let's next talk about a few of those problems that might come up in your new system.

Optimization

There are lots of optimization problems in industry. You have seen several examples already, from projects I was involved with. The first and foremost example is often around the bottleneck in a production chain, but other things have to be optimized as well, for example the placement of pieces on layouts to minimize waste or the calculation of codes for master key systems, as described earlier.

Stock Access

One frequent optimization problem is stock access. I remember a situation they had when I was with the lock manufacturer. When extending the manufacturing facilities, the company built a storage system that would on one hand hold stock with standard locks and accessories and on the other hand all the work in progress as well, all stored in standard boxes. This was to replace an older system holding work in progress only.

The new system was first loaded with all the standard stock items. It had two access slots, and all that stock was stored in the first slot by the system. When gradually switching work in progress over to the new system, all those jobs then went into the second slot.

Because work in progress had way more stock movements than standard stock, once the majority of jobs was stored in the new system, the system started to run into troubles. The second slot was too busy, and the first was almost idle. The company had to redistribute standard stock so that both slots had their fair share of both types of stock and load was evenly balanced.

Another example of stock access optimization is a large meat processor. This manufacturer produces packets of meat products, for example sliced salami, bacon, or ham. The meats are prepared and packed into packets of a certain target weight, so there are packets of 3, 5, and 7 ounces of salami. Weights are not exact, though, so before labeling, the exact weights are determined, the prices are calculated, and the labels are printed.

Customers order electronically. Most customers are large distribution chains, and they order for all their shops. They have their own preprinted labels, which means that those orders are all processed together. Price labels are thus a customer's sale prices and have barcodes as well. Furthermore, customers can specify sales promotions on products, for which a second label is printed.

Goods are stored in a cool room in standard boxes. Every standard box contains a number of packets of a certain target weight. The boxes are stored on long shelves side-by-side. They are accessed by transport shuttles and brought to the front when needed. From there they are transported to the commissioning floor by lifts.

Stock access is the bottleneck of the system and has to be optimized. In the control room are two operators. One does longer-term planning; that is, he lines the jobs to be processed into a queue. The second operator runs the production. When the next batch comes up, he runs the optimizer to calculate optimal storage access sequences. Once the batch has started, he supervises production. He has about ten screens, some of which are just showing critical unsupervised positions in the transport chain.

Once the production sequence of a batch is sorted, the storage system accesses the boxes needed and brings them to the commissioning system. This system consists of several commissioning lines, each containing about 20 boxes into which products are commissioned. At the beginning of each line is a computer controlled by an employee. When a box arrives, the computer tells her how many packets of the product in the box have to go onto her line. The employee counts the packets out and lays them onto the line. The packets then run past the balance and two label printers, are properly priced and labeled, and then go to their respective slot with their target box.

The production sequence is not simply the optimal access sequence for the storage system. The boxes are filled and stapled in dispatch sequence. This means that the box staples can be loaded onto the truck in the correct sequence, so when the truck driver does his tour, the next staple or staples of boxes to unload are always readily accessible at the end of his load.

Furthermore, the boxes are stapled in unloading sequence at the stores as well. Every store has its unloading sequence stored in the meat processors' database. This means an employee at the store can just put the staple of boxes onto a trolley and drive the trolley to the fridges in his store. What is in the top box is what he wants to unload first to his fridge. He can then just go through the row of fridges sequentially, and what he needs next for the fridges is always in the top box.

A third example is the stock feeder load balancing software for a large IGU manufacturing site. Three glass cutters are all accessing the same large stock. Stock is stored on racks in an almost vertical position. A feeder crane picks up stock sheets from those racks and transports them to a tilt table. The tilt table brings the sheets into horizontal position and travels with them to the cutting tables.

Two of the three glass cutters are high-speed cutters; the third is used for laminate glass cutting and is fairly slow. One of the high-speed cutters feeds into a breaking and sorting line. Everything in a standard size range and of rectangular shape runs over this line. This is the runner line and needs to run on high productivity, all running automatically. Feeding that line has priority.

The second high-speed cutter has high priority too. It takes everything non-rectangular, everything that does not target the IGU lines, and everything too big or too small for the sorting system. So, this line has high priority too, but because many things are a bit work intensive to break out, this line works slower.

And as I mentioned before, the laminate cutter runs even slower. However, it has two waiting places in front of it, so they can order early, and the feeder system has time to work out where there is an optimal gap to feed it.

It was clear from the beginning that the feeder would cope with the workload only if the company optimized its processing sequence. The manufacturer provided good data, so I could work out how long it takes the feeder to get a stock sheet from each rack. Because the stock area is so big, these times varied widely, and I had to make some recommendations on where to store the most frequently used stock.

Furthermore, I have to work out how I can look ahead to see which cutter will need glass next. To this end, I dynamically calculate the cutting time for every stock sheet and every cutter. So, if the cutter starts cutting a stock sheet, I know exactly when it will finish, and I can thus figure out which stock sheet to feed next. I can even calculate ahead, knowing the feeder times and the cutting times, to see whether there is room somewhere in the sequence to bring another sheet to the laminate cutter, before it runs out of glass.

To complicate things, there is a rest sheet storing device as well, where the tilt table brings and fetches rest sheets. Bringing rests there costs additional time during which the tilt table is not available. Getting rests from there means the feeder crane is not involved with that feeder job and can thus go to the next one.

Other Optimization Problems

In most cases, there are lots of other local optimization problems that have to be attended to. Because I have discussed glass already, let's have a closer look at the optimization problems in a typical glass processing business.

- Stock access has to be optimized to make sure no cutting line has to wait for glass.
- For every glass type, layouts have to be optimized to minimize waste.
- The processing sequence on the sealing line has to be optimized to ensure continuous operation.

- Packing is organized so that the customer can unpack in the desired sequence but stapled so that transport is safe.
- The distribution tours are organized to minimize transport distances.

Of course, all that packing and dispatch stuff has to respect the laws of physics, the rules of safe transport, and last but not least the sequence wishes of the customers. Distribution tours have to take into account the loading capacity of the trucks, meaning assigning the right trucks and, if necessary, trailers to the right tours and allocating tours so that trucks are not overloaded.

However, after all that local optimization, you might start wondering: does that deliver the overall best result in the end? The more you think about this, the more you come to the conclusion that no, it does not.

Going backward, you will find that the production sequence determines the sequence at which the pieces arrive at the end of the sealing line. If you have worked out an optimal packing sequence, then you might want a completely different sequence to make sure most of the pieces can be packed onto the delivery racks directly and need not be moved around and repacked multiple times.

Further back, the batch sequence has an influence on the sequence in which the pieces are cut. Layouts are sequenced into groups of pieces because the following sorting systems have limited capacity. If all the small pieces are grouped together at the end and all the big pieces come first, then there is a lot of waste on the early stock sheets to cut.

In theory, you could go even further back and try to figure out what effect the cutting sequence has on the feeder crane. But I think there should always be enough capacity there to cater for all situations. However, this example makes a case for needing a more complex sort of optimization if you want to achieve an overall optimum.

Sophisticated User Interfaces

In many mass customization projects, there are complex background computations to be performed to make sure that what the user specifies can actually be produced. This can happen in mere configuration type projects, where you still have to make sure the components picked by the user can be assembled together and the combination makes sense.

This might involve some physics or geometry or ICT know-how, most likely a combination of some if not all of them, such as in the case of a laptop computer from Dell. In many cases, the programming of those rules is done by the engineers who know how to configure these systems, or at least they are involved heavily in the program specifications.

However, in many cases, I have seen programmers unable to understand the complex requirements and engineers unable to understand the programming. In particular, in many parametric mass customization projects geometry data processing is involved. Currently engineers are using computer-aided design (CAD) programs that do all the geometry calculations for them. When automating this process, engineers struggle to specify how those calculations work, and programmers struggle to find the right program structures and methods.

For example, I have designed a user interface for the glass industry where the user can draw an odd shape with the mouse. Once the shape is closed, the user is presented with a set of measurement entries on the drawing. He can then enter all the measurements he knows, and the system automatically calculates all the other ones it can. Measurements are not only shown in the drawing but in a list at the left as well. There it is easy to see whether measurements are missing or whether there is a contradiction (relevant measurements turn red). Users can specify additional measurements (such as diagonals), and with every new measurement, the system tries to complete the shape. If it succeeds, all other measurements are calculated, and the shape is drawn in the correct proportions.

Complex Business Problems

The problems described in the previous section are only a small sample of complex problems that can arise in manufacturing and other businesses. Michalewicz et al. give more examples in their book [28]. They characterize complex business problems as follows:

- The number of possible solutions is too large to evaluate all.
- The problem exists in a time-changing environment.
- The problem is heavily constrained.
- There are many usually conflicting objectives.

In most cases, even optimizing for a single objective, such as to minimize waste or to plan production, has way too many possible solutions to evaluate them all. So, although good algorithms are known for those problems, they usually can't guarantee optimality in all cases. If this is now combined with the last point, the fact that there are many conflicting objectives, then it is already apparent that it can become quite difficult to find an overall optimal solution.

The second constraint means that you have to re-calculate and re-optimize all the time. If you produced the same items all the time, you could come up with a factory layout and a production plan to optimize for producing those items. You could run a supercomputer for a week to find the best solution and then run that for years to come.

In practice, you produce different things every day, so you have to plan again every day. Frequently production does not go to plan, though. There are too many things that can go wrong. Employees call in sick, machines are not working, raw material has not arrived, and so on. That can mean you have to plan production again even throughout the day. So, you don't have all the time in the world to plan optimally.

The third point is harder to understand. The more constrained a problem is, the easier it seems to pick the best of the remaining feasible solutions. Unfortunately, this is not how it works out. The more constrained the problem is, the harder it can become to find the remaining feasible solutions at all. For example, if you can formulate the problem as a linear programming problem, then that can be quite easy to solve. However, if you have to add the constraint that some of the unknowns have to be integers, then this makes the problem highly nonlinear and thus hard to solve.

As many examples have already shown, there is a great need for optimization in the manufacturing industry, and in my experience the proponents of the industry are often unaware of the extent of this need.

This means that most factories do not run at their optimum. Somehow, people are aware of this, and there are always mitigating circumstances to blame. But what I mean is that factories should always run optimally given their actual circumstances.

If you are in an environment with flexible manufacturing systems to deliver customized products, then a lot of things can happen. Customers do not order in your optimal production sequence. Given the circumstances you are in, what is thus the best way to produce what you have to?

In many cases there is simply a bottleneck, and if you can optimize for that, you are pretty much in the clear. In the meat processor example, if you can set the boundaries so that commissioning produces proper sequence and optimized stock access ensures the commissioning lines are not waiting for stock, then you are in the clear. But if this is not always the case, you are paying a price somewhere.

The difficulty then is to figure out what price you are paying in terms of time lost. If you observe this happening every now and then, the first step would be to measure the times of those instances. If the system knows when a box arrives late from stock, it can record those incidents and add up the times.

If those recordings show that you are starting to pay a high price in efficiency, then you have to analyze and consider alternatives. Maybe, if you organize a bit differently, you can eliminate those waiting times and achieve increased speed.

There are many more examples where you can potentially improve overall by significantly improving at one point and paying a small price somewhere else. Often you will also have to improve that other process to ensure the price you pay remains small enough.

One such example was a project I did for a large joinery. Joineries producing individual kitchens tend to build a batch from one job at a time, which might consist of a kitchen and a bathroom or two. Often the materials for those are not even the same, except for the white melamine that is inside everything. This means that the optimization run for the more expensive visible front material is often fairly small and thus tends to have high wastage. The additional setup from the frequent materials changes add to the cost.

Some materials are always "in" and appear in many jobs at the same time. It would thus help if those jobs could be processed together. However, the more jobs that are optimized together, the more there is to be sorted at the end. I thus suggested a setting where only two jobs are optimized together if possible. For each pair of jobs considered, I ran an estimator (which can and will be improved) to figure out the potential savings.

Because this combining of jobs has implications on the schedule, the price to pay is late delivery, particularly if two jobs are combined for which the delivery date is wide apart. Because late delivery has a defined cost, the scheduler then tries to find an optimal solution that balances the two conflicting interests.

The scientific literature is full of examples of such difficult optimization problems and their solutions. For example, in [29], the authors describe a workforce scheduling problem for a highly skilled, highly paid maintenance workforce, where a trade-off has to be made between maintenance cost and machine availability. In [30], the authors describe an algorithm for stockyard planning and machinery scheduling in a coal stockyard, where stockpile planning, stacking, and reclaiming scheduling and product type selection are highly interdependent operations. Z. Michalewicz and D. Fogel give in [31] a good overview over heuristic methods used to solve large instances of these kinds of problems.

Simulation

If you are confronted with a project for a new system with high complexity, the best way to help understand that system is to build a simulation model of it. If you work with one supplier of the whole system or a system integrator that takes responsibility for overall performance, then they might already have something like that.

It will be your responsibility to figure out with the supplier or system integrator what you should be able to produce with that new plant or machinery and what efficiency you expect. Only then will it be possible to simulate those use cases and to find out how you can run those rare but financially interesting jobs efficiently.

If it is not so easy to get hold of a simulation model and you can't find anyone able or willing to write one for you, then it's often tempting to just let it go and hope for the best. Don't try that!

Instead, find a way to get a reasonable simulation done. Talk to your system integrator. They must be interested in doing something like that. First, it will help them to better understand what you need. Second, during the development of their software systems, they can use the simulations of the machines to test their software. This might mean additional effort for them during the development process, but it will be well worth it if the simulation reflects how your plant will perform reasonably well. Your system integrator might be prepared to do something if you don't order anything until you are confident that it will all work well.

If you don't want a system integrator involved at that stage and want to resolve all the specifications yourself before you order (provided you know what you are doing), then you should still try to find someone who can help with a simulation. This could be a third party or even someone in your IT department. There are several tools for discrete event and stochastic process simulations out there. Most likely your system integrator will know the best tool to use. If they can't do that, try to find someone who can handle that.

But please be aware that there are about as many simulations as there are objects to be modeled and simulated. There are classes of similar problems, but there are significant differences as well. Methods working in some cases might perform poorly or completely fail in others. That's why I suggest your system integrator as the first port of call. They should know how your system will work, how to model it, and how to build an appropriate simulation.

If all else fails, have your team run simulations as thought experiments. In any case they will have to come up with use cases they think might be problematic. Let them assess these cases and build a classification so that you know the most critical ones. Have the team carefully think through those cases and manually work through the system to figure out what will happen. I'm almost certain they will find problems, and the longer they work with it, the deeper their understanding of the whole system will be, and that will lead to the changes necessary.

It might well be that some functions are not even helpful, and requirements can be simplified. They might find that in some cases the way certain orders are processed will be quite different from what was initially planned.

For example, in one of my company's most recent automation projects involving IGU production, there were several blocks set up in the sorter to store very thick glass. Thick glass means wide slots and that in turn means fewer slots per storage block. So, even if there were about 80 such slots available, if you have one large order position of, say, 200 units for a façade, then it will not be possible to sort in all the pieces for that position before production begins. Typically those pieces are ordered externally because they are either laminated or toughened.

Traditionally the pieces were loaded onto the sealing line by crane, one after the other, and because that process was slow, the whole process of sealing those units was slow. However, with the new system, there are now two spots to feed those large pieces, one at the sorter and one on the sealing line itself, at a position where feeding from the sorter is not affected. So, with the right software support, it is now possible to sort the first pieces of the units into the sorter and add the second pieces at the sealing line once production starts, thus speeding up the whole process. Although it is not possible in that case to process those orders the standard way, it is a smart workaround that still makes the best possible use of the new machinery.

CHAPTER 11

Commissioning

> *"The shortest and best way to make your fortune is to let people see clearly that it is in their interest to promote yours."*
>
> —Jean de la Bruyère

At the beginning of this book, I promised to come back to the lock and key example. There were two more things I learned from that project, and I will share them with you here.

Back then, it was easy to understand how everything had to work together. Computerizing the system management aspects was straightforward. In contrast, the more intuitive aspect of the solution process, calculating the codes for keys and cylinders to fulfill the specified functionality of the master key system, was not easy. The brute-force approach, trying every combination, would obviously have taken too long. I started reading books on the subject and soon found that my problem belonged to a class of computational problems for which no efficient solution was known. I was confronted with one of those difficult problems I discussed in the previous chapter.

What is more important here is the second lesson I learned from that project. There was a way out of the complex problem. After all, if you could find a solution manually, there should at least be a better one with computer support, and that was in fact what the progressive employees of my department had asked me to do in the first place. When things were done manually, people did not always find the best solution. I understood now that even with the fastest computer in the world I would not be guaranteed to find the best solution either. But all I needed was one that was good enough.

We decided to do both, an automatic solver for the smaller master key systems and a computer-supported one for the large ones. The best coder on my team, a guy who always claimed this would never work, became my biggest fan once the system was up and running. He was in charge of most of the large and complex systems now, and he stated that what used to take him three weeks the manual way he can now do in one day. I learned a big lesson: computers are not there to replace people; they are there to empower them!

Although this was not about automation, this lesson is important in automation projects as well. Of course, you will work with fewer people in the end. The important thing for those fewer people will be, however, that they understand what's happening and what they can affect. So, when talking about putting your new system into service, there are a few issues to be aware of.

The Media Equation

When more and more problems around the human computer interface came up when introducing new IT solutions, I understood that I had to develop a deeper understanding of that area. Thus, as a member of Institute of Electrical and Electronics Engineers (IEEE) and the IEEE Computer Society, I joined the Human Computer Interface special-interest group.

From this, in the 1990s, I came across an interesting book by Byron Reeves and Clifford Nass with the title *The Media Equation* [32]. The authors had proven in a series of experiments that once a representation of something is good enough, we react emotionally to it as if it were another living being. They collected evidence for this in a long series of experiments. The typical approach was to take some known experiment studying the interaction of humans to humans or humans to animals and repeat it, replacing the second being with a computer.

The interesting thing is that it does not take much to trigger the same emotional reaction to the computer as to a living being. In 1999, U.S. market research company Concord Communications found that 83 percent of users hit their computer! Remember those computer rage videos that used to circulate long before YouTube? Just Google *computer rage* and you will find them all!

It is important to understand that the media equation affects everybody, including children and adults, accountants, psychologists, engineers, scientists, programmers, and me and you. There are no exceptions, and our reactions are automatic, which means we have no choice to react differently.

Furthermore, our reactions do not depend a lot on the quality of the interface. There is no significant difference between the high or low resolution of the screen; in fact, a text-only interface will do. This is a reminder that reading a book can make you laugh or cry as well. It is not the picture on the screen we react to; it is the picture in our brain. However, there is a bit of a difference regarding an acoustic interface, in that an unnatural voice is perceived as being less trustworthy; bad synchronicity between image and sound has the same effect.

It does not matter what medium you are dealing with; the media equation holds for all of them. Computers are not simply about efficiency, and television and film are not only about entertainment. With all media, we learn and react, get aroused, or get sad or upset. The reaction to all media is broad and deep.

Size matters as well; a bigger screen or a full-size video on a big screen can convey an unintended message. For example, I will always remember the strange feeling I had when playing Riven. The first time you make it onto that aerial railway, it makes a significant drop, and the fact I did not physically experience what I visually did gave me a sick feeling in my stomach. So, if you try to explain some laws of physics using video from a roller coaster on a large TV or home theater, then your dominant experience is the roller coaster ride. If you are in such an environment and the voiceover tries to explain the physics of your experience, of course that's not what you will remember in the end.

It did not take long for that book to convince me. I have to admit that I did not even read it to the end. However, what surprised me was what people made out of it.

Remember Clippy, the annoying "assistant" in Microsoft Office? Microsoft partially sponsored the research for the book, and I guess Clippy was created as a result of that research. Remember how he used to come up every time you were struggling a bit, just to tell you the obvious stuff that you already knew but never what you needed to know? Of course, Clippy reinforced the perception of the computer as another being and thus was even more annoying. I was always convinced that what we should do instead is emphasize that even with all its complexities, a computer is still nothing but a stupid machine.

Microsoft had a sponsorship deal with the authors of the book that had the condition that Microsoft would get all the research results in advance, before the book was published. Clippy and the other characters were not directly mentioned, but assistants were referenced in the book, so obviously there was some research done. They figured that people would go out of their way to not hurt the assistant's (Clippy's) feelings, for example by turning off the feature (killing Clippy) or changing to another character.

Perception is important, not facts. People know that computers are not really intelligent, and the new machines on the factory floor will not be smart either, but experience counts, not abstract knowledge. Robots used these days to entertain and talk to elderly people in retirement homes are perceived as friendly and funny, and of course these folks know that the machines are only robots. But if they entertain, what's the problem?

Remember that people like simplicity. When faced with trying to understand or build a mental model of a new system, they will go with the simplest model that seems to fit. This means that freedom of choice is not always the preferred solution for the user. Some prefer simplicity even if it means less choice. This can become a challenge in an environment that requests the machinery to handle all of the complexities that can arise from the variety of customer orders.

Chapter 11 | Commissioning

If you are in a large factory hall with all the machines humming and thrumming, then it might be difficult to think of the dozen or so computer screens as all being part of a huge animal. However, people will always try to get their head around how the whole system of interconnected machines works, and having grown up with other humans and animals around us, we tend to use those models when trying to understand.

If you then want to try to make people understand what that system does, what exceptions can arise in each worker's environment, what the exceptions mean, and what the worker can do about the problem, then you always have to emphasize that the whole system is only a set of interconnected stupid machines. That way, you will come to a state where the workers are able to intervene in a meaningful way.

Ease of use can also be achieved by trying to implement functionality in a way that comes naturally to people. Trying to follow the rules of the social and natural world can help to make user interfaces more intuitive and reduce the need for manuals. An active model of the factory layout (or a relevant segment) can help a lot in illustrating what's happening. If users can click critical points and intervene directly, then that provides an easy-to-understand interface.

Reading and Literacy

When changing from an environment with little automation to one with a high degree of automation, all of a sudden computers are everywhere on the factory floor, and people are forced to use them. Somehow, the old problems of the 1970s and 1980s, when computers first came to the top floor, seem to reemerge. However, this time there are a few additional problems.

To get the problems out of the way, let's talk about the old problems first. The first problem encountered by just using the old alphanumeric screens was related to the fact that as people get older, they lose their ability to focus over a short distance; in other words, older people need reading glasses. Sometimes people did not want to talk about being unable to read the screen, so they just rejected that new system as a whole. When I implemented a workstation solution once in the early 1990s, I convinced an elderly employee that he could use the new system simply by increasing the font size!

Apart from the fact that workers older than 50 are as vain as anybody, sometimes reading glasses on the work floor can be a pain. Someone might even have reading glasses at home, but that doesn't help if someday someone comes and wants to do some training on a new system. They simply can't see, and there is no quick way out.

Mass Customization

I'm short-sighted in one eye and close to normal in the other. This means I do not need glasses in everyday life despite being well over 60, although I use glasses to drive a car. Peter (not his real name), a development partner with one of my customers, is really tall, about six inches higher than me. Using glasses he sees well, but he has a back problem and thus uses a desk where he can stand. He has placed his screen at the rear edge of the desk so he has lots of room on the desk and can still easily read the screen.

When we tried to sort out an issue together, I tried to look at his screen. However, with the desk so high up, I could not lean forward and over it and thus had no chance of reading what was on the screen. It did cost me a bit of an effort to tell Peter that I had a problem and that we should maybe find a solution that catered for both our disabilities—his back and my eyes. After that I had a bit more compassion and understanding for that elderly guy some 20 years earlier.

Another problem came later, when screens became colored. About 7 percent of males, or about 1 in 14, have a form of red or red-green color blindness. These conditions are on the X chromosomes, and that's why they are much less common in females. Most people don't know about their condition themselves. Errors and misunderstandings can easily result.

While in the western world basic literacy skills are usually greater than 95 percent, this means that more than 95 percent of people are able to read simple text on a familiar subject. If it comes to reading more complex things, such as the instruction leaflets for medicine, then typically one person in six can't understand the text.

In countries with high immigration rates of foreign-language speakers, such instructions usually come in the most commonly used languages in that country. If you think about a computer program, the situation is even worse. Although most of our programs "talk" two or three languages, the only language in which those programs communicate with the outside world is the language of the operating system, which in turn for most installations will be the dominant language of the place where it is installed.

This can mean that to the about 17 percent of people with a reading comprehension problem, there is an additional percentage of people who, while being able to read in their native language, cannot comprehend very well the language the computer uses.

I'm not aware of a country in the western world with an unemployment rate of 20 percent or more. So, where do you think those people with limited literacy skills are hiding? On your factory floor! And of course those people are not proud of their lack of reading comprehension. They will try to hide this fact.

Almost 20 years ago my company installed some software in a factory, and I was approached by the production manager, concerned about exactly this literacy problem. He was one of the few I've encountered who was aware of it.

The story he told me was that the company had recently implemented some minor changes to the production papers printed by their mainframe. What he didn't realize was that many people just knew that they had to type in the number found at a specific place on those papers before starting with the job. However, because some of the fields had been moved around (with their labels!), some people could no longer find the correct data and entered the wrong numbers. Until they discovered this issue, quite some time after the change, their production was a mess.

Summary

To sum it up, you have to prepare your employees for the new jobs. You have to make sure they know what to do at their place and what's happening around them. You have to find out who can cope with which roles and where to place your people best.

There will be a lot of hiccups when you start a complex new system. I've seen errors in programmable logic controller (PLC) programs that led to machines destroying themselves. Common problems are incomplete adjustments, improperly tightened screws leading to sensors moving out of place, overtightened belt stretchers leading to the belts breaking, and similar things. And of course there will be software bugs, because that is the most complex part of the system.

This means you can't just flick a switch to stop the old system and start the new one. There will be a changeover period whose length is mainly determined by the complexity of the new system and the thoroughness of your preparation. For that period of time you have to make sure you can still deliver to your customers.

So, make sure you come through this phase well. Try to keep as much production capacity in the old system as you can. Maybe temporarily change to two or three shifts there to make sure you come through the transition in good shape. You might even need more capacity in the transition phase, even though you might have planned the transition for a time where order volume is low. This is because as soon as the new system starts running, you will need some of your best people there. This gives the new system the best chance to get working and your best workers the chance to learn the new system. However, the new system will not run smoothly, so your best people will hang around and wait for the system to get running again for quite some time. I call this effect the *investment speed bump*, as shown in Figure 11-1.

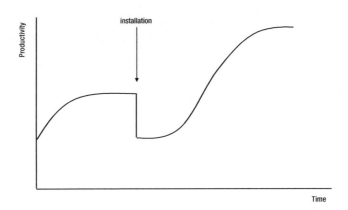

Figure 11-1. Investment speed bump

You can prepare yourself and the team well to make the investment speed bump as small as possible, but you can't completely avoid it. What you can avoid with good preparation, though, is the speed bump heavily affecting your overall production.

Once the new system is up and running well, do not immediately fire excess people. For some time at least, find work for those who are no longer needed in the new operation and for those who despite of all your efforts can't cope with the new work environment. Find some other work for them. After all, while you were struggling to get the new system running, they were there and made sure production was kept up.

If you fire them now, that has a devastating effect on the rest of the workforce. They will feel betrayed. After all, it was the whole team that made the new system work. So, instead of firing people, try to find the right place for everybody, and if someone becomes really useless, wait until that is apparent for all to see.

If the main purpose of your investment was to increase production, be it by better customization, better products, shorter delivery times, or a combination of these and other improvements, then your production volume should increase anyway. This could mean you don't have to reduce staff at all, or at least to a lesser extent, to achieve your investment goal.

CHAPTER 12

Implementation Process

"The world is moving so fast these days that the man who says it can't be done is generally interrupted by somebody doing it."

—Elbert Hubbard

Let's assume you have developed an idea of how you can make mass customization work in your business. To make it happen, you have to communicate this vision, and you have to form an implementation team.

It can be a long way from an idea to a vision, a team, and a detailed implementation plan, and depending on your position in the company, there will be different ways to get there. But wherever you are in the company, there will be some convincing work to do. So, let's talk about those things first and then talk about the implementation process.

Building the Vision

Once you have an idea of where you want to go, you have to build a team that can come up with a clear vision for the future of your business, and then you communicate and inspire the majority of the employees involved to make it happen. Having that vision is nominally the task of the board. Making it happen is the job of the executive team. But you know that this is not how it works.

Chapter 12 | Implementation Process

For your ideas to become reality, the executive team has to be convinced. More than that, they have to adopt the idea. The best thing that can happen is when they think it was their own idea in the first place. If you think about it, to a large extent this will in fact be the case anyway. If you come with a big idea somewhere high up in the clouds, you might not even know yourself how much of that can be realized and in what timeframe.

So, let the team work on it. Even if you are the chief executive officer (CEO) or a board member and have worked out most details of your vision, don't tell them in every detail what it is and what you want to achieve in what time frame. Ask them instead what they think of your idea and whether they would be able to make something in that direction work. Let them think about it and meet for further discussions later. Chances are, they will develop their own ideas too, and some of them might even be better than yours.

If you are not that high up in the hierarchy, you will have some convincing work to do first anyway. Discuss the idea among your peers. Try to convince the more progressive ones early and let them help you to flesh out the vision, before the more conservative people start to claim that this will never work and try to kill the process in its infancy.

To build the vision, think about all your products. Where do you want to start? Are there opportunities to extend the production volume if you could mass customize some of your products? Are there other products you could easily produce using mass customization with your technologies, your machines, and maybe some additional investment? It does not have to be fully automated from the start. Maybe it never will. The main thing is it becomes competitive and profitable.

Try to understand how this could grow. Maybe you could extend mass customization into most of your products over time. Or you see more opportunities with more and better products. Plan where you want to be in two years and in five years. And be aware that these goals will change. So, try to stay flexible in every investment phase you plan. Work out how you can be agile and how you can proceed in steps, and remain flexible.

Assembling the Team

Who is the team? I bet you already know them. It's those people who want to move things, those who always try to change and improve, and those who always push although it might not really be their job to do so.

To me, building the team and building the vision are two processes that can't be separated, although not everybody that helped with building the vision will be on the team, and not everyone on the team will be helping to build the vision.

Of course, the team has to include people who understand the processes and systems involved. You must involve people who understand the technologies you use. As you saw earlier, this is not necessarily the production manager, although I would not exclude him outright, especially if he belongs to the category of people described earlier. If you want to go anywhere near what I have described in this book so far, then you will have to involve people from IT.

If you are involved with a large company that has a professional team that does all the automation projects, then first check how they perform. If they still operate the way they used to some 20 years ago, then they might not be part of your team. On the other hand, there might be people in that team who sense that something has changed and want to work differently. I'm sure you will find them if they exist.

You might want to involve external partners as well, if you work with them a lot and can trust them fully. The only problem here is if they work with the competition too (for example software houses providing software specific for your industry); then you might not be able to trust them *not* pass on inside information to your competitors.

If you want to be agile, then the discussion will automatically widen. Remaining flexible while becoming more efficient means that to some extent you have to know where the business might go. So, the discussion will have to be about future products and how they could be produced and mass customized. Of course, the discussion will first be about mass customizing existing products and what would have to change to make this happen.

The next step will be to come up with a plan for the investments you need to make and when, what you want to achieve with each phase, and some cost estimates. Once the team has signed off on the plan, you will plan in more detail and implement the first step.

First find the people in-house who you think are best suited to make a start. I discussed before how you can find them. This team might not yet be complete, but at least they can make a start. Don't just confront them with your vision. Discuss ideas, strengths and weaknesses, opportunities, and threats. Listen to their ideas, and extend and adjust your vision where needed.

The skills you need on the team but can't find in-house you will have to find externally. You might already work with external people in some functions, or you know the right people in your network. Maybe someone on the current team knows someone in their network. Bring them in.

Communicate the Vision

So, once you have the team and the vision, you will have to inform and inspire the wider community in your business. The bigger the vision and the more it will change your business over time, the harder it is to explain and convince people. There are whole books written about those processes, and I don't want to go into details here except by recommending you read Scott Keller and Colin Price's *Beyond Performance* [33].

When it's time to communicate your plans to the employees, to all other stakeholders, and to the wider community, this should be done by the highest-ranking officer and, if possible, delivered first to the whole workforce. Not everybody will be convinced from the start. Depending on how much behavioral change you need from your people, it will be necessary to make sure they are convinced and on board. This might take a bit of time and require all the team members to do some convincing work. See [33] for examples of how to do this.

Iteration

As suggested throughout the book, it is optimal if you can implement your vision in multiple iterations. Try to plan the iterations so that you do the most important things first and so that you get the best return on investment (ROI) early. Each iteration goes through several phases as follows.

Assessment

In the assessment period, you do the following:

- Work out where you are currently with your manufacturing processes. After a few iterations, this will be little work.
- Work out where you want to go with this iteration.
- Work out what you need (for example, in terms of machinery and software) to achieve this.

Design

In the design period, you do the following:

- Work out the project details with machinery suppliers, your internal and external IT providers, and where applicable your systems integrator.
- Make sure all parties involved are clear about your goals and committed to do their part to achieve them.
- Plan all necessary details.
- Build a model and run simulations.
- Assess simulation results and modify your plans where necessary.

Implement

During the implementation, you do the following:

- Specify and order all the machines.
- Configure and where necessary develop all software needed.
- Make sure all interfaces are appropriate and well understood. This concerns user interfaces as well as software interfaces to machines and between software products.
- Plan and implement training of all employees concerned with this iteration.
- Plan the installation phases with the machine and software providers.

Install and Test

In the install and test portion, you do the following:

- Be prepared for this phase and mind the investment speed bump!
- Execute the planned installation phases and test as early as possible.
- Test the first small production runs on the new system.
- Let the users work with the new system during those first production runs so they can familiarize themselves with it.
- Ramp up production.

Maintenance and Support

In the maintenance and support portion, you do the following:

- Resolve remaining issues and problems.
- Get your software provider to analyze logs to identify issues.
- Establish preventative maintenance.
- Identify changes and adjustments needed for changed requirements or market demand.

Re-iterate as often as needed.

Mass customization can and will increase your production, but it demands careful implementation. Take the lessons of this book to heart, but don't be afraid to go for it. Engage with your employees, your machinery supplier, and your systems integrator to make sure that your investment funds are well spent—improving your market position and better serving your customers through mass customization.

Continuous Improvements

Once you've understood your problem, what you want to do about it, what areas to emphasize, and where to begin investing and once you have gained overall clarity, you are ready to make your investment decision.

If you have managed to follow my most important recommendations so far, then your first investments will cover the most important things you need in order to improve your production now. And if you have invested wisely, then you will have some money left for improvements later.

Having understood the problem well means you also know what possible extensions you could invest in later, and you made sure these extensions will be easy to integrate with what you have invested so far.

Now you have the money and the flexibility to invest where you have the highest need and where you can get the best return on investment. If market conditions change, if you have new products or your product mix changes, then it is much easier to react, not in the least because you have not spent all your money on the initial project and tried to cover all eventualities there, even those you did not know about at the time.

Once your new system is up and running, there will be many issues. You can try to capture the smaller practical issues with some form of reporting system, or your supporters might collect and report them all. Some issues might be within warranty, for example something not working as specified, and others might be bigger and more complex. You could use the critical incident technique to find the smaller issues, and the bigger ones will make themselves manifest.

Earlier I talked about issues that elderly workers might encounter in your new factory. If you can't change your new system and you did not foresee the problem, you may needlessly lose workers with vast experience. If they cannot work efficiently in their new workplace, then you will have to find them a new job.

Since in your new factory, manufacturing is no longer solely about production—you have automated that—there are a lot of opportunities in other fields for skilled workers. These jobs can be in maintenance or service delivery, for example giving advice on the shop floor or support over the phone or onsite.

A flexible retirement system, especially for highly skilled employees, can also help to facilitate knowledge transfer to the younger generation. Many times I've seen that it becomes apparent only after some employees have left that some valuable know-how was lost. It can then become a painful exercise to rebuild and reconstruct from scratch.

Another example for better integration of older workers is BMW, which redesigned their assembly lines to support older workers [34]. To their own surprise, they found that younger workers working on those lines had reduced injuries and sickness as well.

But there is more to continuous improvements than just the first year of running your new plant. Business has become very dynamic. There are new products, new customers, a new product mix, varying volumes, changes in the market, and changes on the supply side. You have to be prepared for that type of change.

If your system is flexible and your software is written well, then your company might be able to adapt to a wide range of changes. In particular, software can be written so that it develops some sense of change and adapts accordingly. However, this is at best limited to your software running the system. If more than that has to change, for example if you need new machinery for new products or if a product with low volume becomes a best seller and you can't cope with the demand, then you have to go back to the drawing board.

Remember your simulation model? That comes in handy now. All you have to do is add the changes, build your new use cases, and run it, modify it, play with it until it works properly. So again, it comes down to what I said earlier in the book: you have to be agile. Don't try to come up with an automation solution that is perfect and solves all your problems. You will never get there because by the time you are there, something has changed in the market and your target has shifted. Better stick with that good old rule to invest where you have the best return on investment, and invest only to an extent that allows you to invest more in coming years. Because you will have to.

CHAPTER

13

References

[1] H. Kagermann, W. Wahlster, J. Helbig, "Umsetzungsempfehlungen für das Zukunftsprojekt Industrie 4.0. Abschlussbericht des Arbeitskreises Industrie 4.0," April 2013, accessed February 16, 2015. www.plattform-i40.de/umsetzungsempfehlungen-f%C3%BCr-das-zukunftsprojekt-industrie-40-0.

[2] VDE Association for Electrical, Electronic & Information Technologies, "The German Standardization Roadmap Industrie 4.0," November 2013, accessed February 16, 2015. https://www.dke.de/de/std/Documents/RZ_RoadMap%20Industrie%204-0_engl_web.pdf.

[3] A. Johnson, "What is the Fourth Industrial Revolution," *Manufacturer's Monthly*, April 23, 2014, accessed February 16, 2015, www.manmonthly.com.au/features/what-is-the-fourth-industrial-revolution.

[4] G. Stepank, *Software Project Secrets. Why Software Projects Fail* (Berkeley: Apress, 2005).

[5] E. Dijkstra, "On the nature of Computing Science," August 10, 1984, accessed February 16, 2015. www.cs.utexas.edu/users/EWD/transcriptions/EWD08xx/EWD896.html.

[6] G. Roos, Presentation at the Manufacturing the Future event organized by the Bankstown Business Advisory Service on April 2, 2014, accessed February 16, 2015. https://www.youtube.com/watch?v=oHbAg1BJWP8.

[7] J. Pepitone, "Jetsons Age Beckons, But Opinion Split on Driverless Cars, Drones," NBC News Digital, February 12, 2014, accessed February 16, 2015. www.nbcnews.com/tech/innovation/jetsons-age-beckons-opinion-split-driverless-car-drones-n27221.

Chapter 13 | References

[8] H. Xie, T. Gu, X. Tao, and J. Lu: "MaLoc: A Practical Magnetic Fingerprint Approach to Indoor Localisation using smartphones" (Proceedings of the 2014 ACM International Joint Conference on Pervasive and Ubiquitous Computing, Seattle, Washington, September 13–17, 2014).

[9] J. Armstrong, Y. A. Sekercioglu, and A. Neild, "Visible Light Positioning: A Roadmap for International Standardization," *IEEE Communications Magazine*, December 2013.

[10] +"Need for a Trillion Sensors Roadmap", www.tsensorssummit.org/Resources/Why%20TSensors%20Roadmap.pdf, accessed February 19, 2015.

[11] "Ericson and LEGO at MWC2012", www.ericsson.com/uxblog/2012/05/ericsson-and-lego-at-mwc-2012/, accessed February 19, 2015.

[12] D. Parsonson, "Benefits of 3D Printing for patterns, mandrels and moulds," accessed February 16, 2015. www.compositesaustralia.com.au/2014-composites-australia-and-crc-acs-conference-presentations/.

[13] "Professor Roy Green shares a manufacturing success story", www.youtube.com/watch?v=rSMOmRxS9rg&feature=youtu.be, accessed February 19, 2015.

[14] "Composites Conference 2014 – Newcastle, NSW" www.compositesaustralia.com.au/event/2014conference/, accessed February 19, 2015.

[15] "Program Fact Sheet" www.boeing.com/boeing/commercial/787family/programfacts.page, accessed February 19, 2015.

[16] A. Beehag, G. Prusty, G. Pearce, J. Bradshaw, and J. Pearson, "Development of automated composite research facility at University of New South Wales," accessed on February 16, 2015. www.compositesaustralia.com.au/2014-composites-australia-and-crc-acs-conference-presentations/.

[17] C. Duty "Carbon Fiber Reinforced Polymer Additive Manufacturing" accessed on February 28, 2015. www.cfcomposites.org/PDF/Carbon%20Fiber%20Consortium%20-%20CFRP%20Duty.pdf

[18] J. Khan, "Welcome to the world of Nanotechnology," *National Geographic*, June 2006.

[19] R. V. Gorbachev, J. C. W. Song, G. L. Yu, A. V. Kretinin, F. Withers, Y. Cao, A. Mishchenko, I. V. Grigorieva, K. S. Novoselov, L. S. Levitov, A. K. Geim, "Detecting topological current in graphene superlattices," Science Vol 346 no. 6208 pp. 448–451, October 24, 2014.

[20] E. Brewer, "Towards Robust Distributed Systems," (Proceedings of the 19th Annual ACM Symposium on the Principles of Distributed Computing, 2000).

[21] S. Gilbert and N. Lynch, "Brewer's conjecture and the feasibility of consistent, available, partition-tolerant web services," *ACM SIGACT News*, Volume 33 Issue 2 (2002), 51–59.

[22] E. Brewer, "CAP Twelve Years Later: How the Rules have Changed," *IEEE Computer*, February 2012.

[23] A. Nusca, "Picking the brain of 'rock star CIO' Vivek Kundra," *CXO*, April 24, 2013.

[24] S. Hanoun, D. Creighton, S. Nahavandi, and H. Kull. "Solving a Multiobjective Job Shop Scheduling Problem using Pareto Archived Cuckoo Search," (17th IEEE International Conference on Emerging Technologies and Factory Automation, Krakow Poland, September 17–21, 2012).

[25] R. E. Bixby, "The Latest Advances in Mixed-Integer Programming Solvers," 2008, accessed February 16, 2015. https://symposia.cirrelt.ca/system/documents/0000/0136/Bixby.pdf.

[26] A. Regalado, "EmTech: Illumina Says 228,000 Human Genomes Will Be Sequenced This Year," *MIT Technology Review*, September 24, 2014, accessed on February 16, 2015, www.technologyreview.com/news/531091/emtech-illumina-says-228000-human-genomes-will-be-sequenced-this-year/.

[27] A.M. Turing, "On Computable Numbers, with an Application to the Entscheidungsproblem" (Proceedings of the London Mathematical Society, 1937).

[28] Z. Michalewicz, M. Schmidt, M. Michalewicz, and C. Chiriac, *Adaptive Business Intelligence* (Berlin: Springer, 2010).

[29] N. Safaei, D. Banjevic, and A. Jardine: "Multi-objective Simulated Annealing for a Maintenance Workforce Scheduling Problem: A Case Study," Chapter 2 in *Simulated Annealing*, edited by Cher Ming Tan, InTech Publishing Group, 2008. Available from www.intechopen.com/books/simulated_annealing/multi-objective_simulated_annealing_for_a_maintenance_workforce_scheduling_problem__a_case_study, accessed February 19, 2015.

[30] S. Hanoun, B. Khan, M. Johnstone, S. Nahavandi, and D. Creighton, "An Effective Heuristic for Stockyard Planning and Machinery Scheduling at a Coal Handling Facility" (Proceedings of the IEEE 2013 International Conference on Industrial Informatics, July 29–31, 2013, Bochum, Germany).

[31] Z. Michalewicz and D.B. Fogel, *How to solve It: Modern Heuristics*, Second Edition (Berlin: Springer, 2010).

[32] B. Reeves and C. Nass, "The Media Equation" (Cambridge: Cambridge University, 1996).

[33] S. Keller and C. Price, *Beyond Performance* (Hoboken, New Jersey: John Wiley & Sons, 2011).

[34] M. Flinn, "'Grey' workers hold the key to manufacturing's future," *Manufacturers Monthly*, March 4, 2014.

Index

A

Additive manufacturing technologies, 63
Assistive automation, 66
    Automation projects. *See also* Complex automation projects
    abstract, 24
    automating repetitive work, 30
    customization project, 28
    design construction, 31
    documentation, 28
    domain, 30
    projects fail, 33
    requirements, 25
    research, 30
    risk mitigation strategy, 29
    software complexity, 23
    software construction, 31
    technologies, 28
    user interface, 31

B

Brainwriter, 57
Broadband infrastructure, 37
Business-to-business (B2B), 5, 75
Byzantine generals problems, 79

C

CAP theorem
    Amazon, 82
    bank ATM networks, 83
    computer clusters, 81
    consistency, 84
    eventual consistency, 82
    issue of computer, 83
    network problems, 81
    properties, 82
Carbon-based nanotechnology, 69
Carbon fiber, 67
Challenges
    broadband infrastructure, 37
    complexity, 36
    framework, 39
    resource efficient, 39
    security expert, 38
    standardization, 35
    training and professional development, 38
    work organization and design, 38
Chief information officer (CIO), 25
Cloud applications, 75
Cloud system
    Byzantine generals problems, 79
    CAP theorem, 81
    chain management systems, 86
    computing power, 77
    customer relationship management, 78
    customer service, 85
    data entry, 85
    devices, 78
    economies, 77
    extended cloud, 84
    Internet access, 86
    kitchen and web site, 88
    manufacturing context, 85
    small glass processor, 87

Cloud system (cont.)
  suppliers, 86
  two-phase commit protocol, 80
Commissioning
  automation, 110
  lock and key, 109
  media equation, 110
  reading and literacy, 112
Commonwealth Scientific and Industrial Research Organisation (CSIRO), 73
Complex automation projects, 22
Complex business problems
  algorithms, 104
  characterize, 104
  constraint, 104
  feasible solutions, 105
  kitchens, 106
  scientific literature, 106
Complexity, 36
Composites, 67
Computer-aided design (CAD) programs, 104
Customer relationship management (CRM), 6, 78
Customization project, 28
Cyber physical production system (CPPS), 4

D

Direct digital manufacturing (DDM)
  manufacturing firm, 1
  mass customization projects, 2
  product design, 1

E

Electroencephalography (EEG), 57
Enterprise resource planning (ERP) system, 78
Eventual consistency, 82

F

Factory floor interface devices, 55
  brainwriter, 57
  gesture recognition, 56
  Google Glass, 57

G, H

G-code, 36
Gesture recognition, 56
Google Glass, 57
Graphene, 69

I, J, K

Implementation process
  assessment period, 120
  communicate, 120
  design period, 121
  implementation, 121
  improvements, 122
  install and test portion, 121
  iterations, 120
  maintenance and support, 122
  team assemble, 118
  vision, 117
Impossible Labs. *See* Brainwriter
Information and communications technology (ICT), 25, 37
  business-to-business, 5
  customers and suppliers, 6
  data processing, 3
  ever-changing environment, 3
  manufacturing-related technologies, 6
  mass customization, 2, 4–5
  MES, 4
  new investment speed bump, 6
  Raspberry Pi, 46
  roles in, 3
  self-driving cars, 51
  smartphones and tablets, 45
  systems integrator, 6
  3D printing technology, 47
  timber stairs manufacturer, 2
  visible light communication and positioning, 48
Institute of Electrical and Electronics Engineers (IEEE), 110
Insulation glass units (IGUs), 50, 84
  car manufacturers, 11
  glass cutter, 11
  glass manufacturer, 9
  materials, Industry 4.0, 11

Index

production level, 11
window manufacturing, 9–10
Internet of Anything (IoA), 60
Internet of Everything (IoE), 60
Internet of Things (IoT), 12, 35
   axis controllers, 62
   devices communicate, 61
   GPS system, 60
   ideal devices, 61
   IoE and IoA, 60
   MYOSD web site, 62
   RFID tags, 60
   wireless communication, 61

L

Light-emitting diodes (LEDs), 48
Literacy, 113
Lock cylinder, 16

M

Manufacturing execution system (MES), 4, 12, 14
Manufacturing technologies
   additive manufacturing, 63
   assistive automation, 66
   case studies
      locks and keys, 16
      processing glass, 18
   composites, 67
   computer and software technologies, 59
   Industry 4.0
      aspect of, 12
      electronic system, 13
      IoT, 12
      manufacturing execution system, 12, 14
      production method, 14
      RFID tag, 13
      Wi-Fi net, 14
   mass customization types, 15
   nanotechnology, 68
   robotics, 64
   smart factory
      definition, 9
      IGUs. Insulation glass units (IGUs)

Manufacturing technologies. See Internet of Things (IoT)
Mass customization projects, 103
Master key management software, 84
Master key system, 16
Mathematics
   math matters, 98
   need of, 98
   optimization problems, 100, 102
   stock access, 100
Math matters, 98
Media equation, 110
Monocrystalline form, 68

N

Nanotechnology, 68

O

Optimizing system performance, 93
   complex business problems, 104
   complex engineering problems
      complexity, 94
      Lean Six Sigma, 93
      mathematics, 98
      problems, 95
   extension process, 97
   glass applications, 96
   insulation glass units, 96
   kitchen manufacturers, 96
   production, 95
   simulation, 106
   user interface, 104

P

Programmable logic controllers (PLCs), 22

Q

Quick response (QR) codes, 46

R

Radio-frequency identification (RFID) tag, 13, 48, 60
Raspberry Pi, 46

Index

References, 125
Resource efficient, 39
Return on investment (ROI), 63, 120
Risk mitigation strategy, 29
Robotics, 64

S

Self-driving cars, 51
Simulation model, 106, 123
Small and medium-sized enterprises (SMEs), 66, 73
Smart factory projects
    complex automation projects, 22
    projects fail, 21
    right methods, 22
Smart factory projects. *See* Automation projects
Smartphones and tablets, 45
Software complexity, 23
Software integration
    B2B customers, 75
    cloud applications, 75
    customer, 74
    database, 74
    individualized production, 74
    inefficient data processing, 74
    order-processing system, 73
    products/modules, 74
    small and medium-sized enterprises, 73
    supplier's pricing model, 75
Standardization, 35
Stock access, 100
Subject matter experts (SMEs), 36

T

Teeth-straightening system, 64
3D printing technology, 47
Training and professional development, 38
Two-phase commit protocol, 80

U

User interface, 104

V

Vehicle-to-vehicle communication (V2V), 51
Visible light communication (VLC)
    advantages and disadvantages of LiFi, 49
    GPS satellites, 50
    IGU, 50
    industrial environment, 49
    LEDs, 48–49
    pilot application, 50
    RFID, 48
Visible light positioning (VLP). *See* Visible light communication (VLC)

W, X, Y, Z

Work organization and design, 38

Get the eBook for only $10!

Now you can take the weightless companion with you anywhere, anytime. Your purchase of this book entitles you to 3 electronic versions for only $10.

This Apress title will prove so indispensible that you'll want to carry it with you everywhere, which is why we are offering the eBook in 3 formats for only $10 if you have already purchased the print book.

Convenient and fully searchable, the PDF version enables you to easily find and copy code—or perform examples by quickly toggling between instructions and applications. The MOBI format is ideal for your Kindle, while the ePUB can be utilized on a variety of mobile devices.

Go to www.apress.com/promo/tendollars to purchase your companion eBook.

All Apress eBooks are subject to copyright. All rights are reserved by the Publisher, whether the whole or part of the material is concerned, specifically the rights of translation, reprinting, reuse of illustrations, recitation, broadcasting, reproduction on microfilms or in any other physical way, and transmission or information storage and retrieval, electronic adaptation, computer software, or by similar or dissimilar methodology now known or hereafter developed. Exempted from this legal reservation are brief excerpts in connection with reviews or scholarly analysis or material supplied specifically for the purpose of being entered and executed on a computer system, for exclusive use by the purchaser of the work. Duplication of this publication or parts thereof is permitted only under the provisions of the Copyright Law of the Publisher's location, in its current version, and permission for use must always be obtained from Springer. Permissions for use may be obtained through RightsLink at the Copyright Clearance Center. Violations are liable to prosecution under the respective Copyright Law.

Other Apress Titles You Will Find Useful

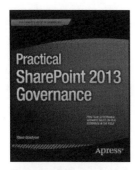

Practical SharePoint 2013 Governance
Goodyear
978-1-4302-4887-3

Sensor Technologies
McGrath
978-1-4302-6013-4

The Profitable Supply Chain
Ganesan
978-1-4842-0527-3

How to Uncover Corporate Fraud
Bell
978-1-4842-0911-0

Disruption by Design
Paetz
978-1-4302-4632-9

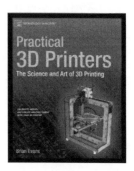

Practical 3D Printers
Evans
978-1-4302-4392-2

Inventors at Work
Stern
978-1-4302-4506-3

Supplier Relationship Management
Easton/Hales/Schuh/Triplat
978-1-4302-6259-6

The Coder's Path to Wealth and Independence
Beckner
978-1-4842-0422-1

Available at www.apress.com

Printed in Great Britain
by Amazon.co.uk, Ltd.,
Marston Gate.